Vol. 27. **Rock and Mineral Analysis.** *Second Edition.* By Wesley M. Maxwell

Vol. 28. **The Analytical Chemistry of Nitrogen and Its Compounds** (*in two parts*). Edited by C. A. Streuli and Philip R. Averell

Vol. 29. **The Analytical Chemistry of Sulfur and Its Compounds** (*in three parts*). By J. H. Karchmer

Vol. 30. **Ultramicro Elemental Analysis.** By. Günther Tölg

Vol. 31. **Photometric Organic Analysis** (*in two parts*). By Eugene Sawicki

Vol. 32. **Determination of Organic Compounds: Methods and Procedures.** By Frederick T. Weiss

Vol. 33. **Masking and Demasking of Chemical Reactions.** By D. D. Perrin

Vol. 34. **Neutron Activation Analysis.** By D. De Soete, R. Gijbels, and J. Hoste

Vol. 35. **Laser Raman Spectroscopy.** By Marvin C. Tobin

Vol. 36. **Emission Spectrochemical Analysis.** By Morris Slavin

Vol. 37. **Analytical Chemistry of Phosphorus Compounds.** Edited by M. Halmann

Vol. 38. **Luminescence Spectroscopy in Analytical Chemistry.** By J. D. Winefordner, S. G. Schulman and T. C. O'Haver

Vol. 39. **Activation Analysis with Neutron Generators.** By Sam S. Nargolwalla and Edwin P. Przybylowicz

Vol. 40. **Determination of Gaseous Elements in Metals.** Edited by Lynn L. Lewis, Laben M. Melnick, and Ben D. Holt

Vol. 41. **Analysis of Silicones.** Edited by A. Lee Smith

Vol. 42. **Foundations of Ultracentrifugal Analysis.** By H. Fujita

Vol. 43. **Chemical Infrared Fourier Transform Spectroscopy.** By Peter R. Griffiths

Vol. 44. **Microscale Manipulators in Chemistry.** By T. S. Ma and V. Horak

Vol. 45. **Thermometric Titrations.** By J. Barthel

Vol. 46. **Trace Analysis: Spectroscopic Methods for Elements.** Edited by J. D. Winefordner

Vol. 47. **Contamination Control in Trace Element Analysis.** By Morris Zief and James W. Mitchell

Vol. 48. **Analytical Applications of NMR.** By D. E. Leyden and R. H. Cox

Vol. 49. **Measurement of Dissolved Oxygen.** By Michael L. Hitchman

Vol. 50. **Analytical Laser Spectroscopy.** Edited by Nicolo Omenetto

Vol. 51. **Trace Element Analysis of Geological Materials.** By Roger D. Reeves and Robert R. Brooks

Vol. 52. **Chemical Analysis by Microwave Rotational Spectroscopy.** By Ravi Varma and Lawrence W. Hrubesh

Vol. 53. **Information Theory As Applied to Chemical Analysis.** By Karel Eckschlager and Vladimir Štěpánek

Vol. 54. **Applied Infrared Spectroscopy: Fundamentsls, Techniques, and Analytical Problem-solving.** By A. Lee Smith

Vol. 55. **Archaeological Chemistry.** By Zvi Goffer

Vol. 56. **Immobilized Enzymes in Analytical and Clinical Chemistry.** By P. W. Carr and L. D. Bowers

Vol. 57. **Photoacoustics and Photoacoustic Spectroscopy.** By Allan Rosencwaig

Vol. 58. **Analysis of Pesticide Residues.** Edited by Anson Moye

Vol. 59. **Affinity Chromatography.** By William H. Scouten

Vol. 60. **Quality Control in Analytical Chemistry.** By G. Kateman and F. W. Pijpers

Vol. 61. **Direct Characterization of Fineparticles.** By Brian H. Kaye

Flame Chemiluminescence Analysis by Molecular Emission Cavity Detection

CHEMICAL ANALYSIS

A SERIES OF MONOGRAPHS ON
ANALYTICAL CHEMISTRY AND ITS APPLICATION

Editor
J. D. WINEFORDNER
Editor Emeritus: **I. M. KOLTHOFF**

Advisory Board

Fred W. Billmeyer, Jr Victor G. Mossotti
Eli Grushka A. Lee Smith
Barry L. Karger Bernard Tremillon
Viliam Krivan T. S. West

VOLUME 129

WILEY

JOHN WILEY & SONS
Chichester / New York / Brisbane / Toronto / Singapore

Flame Chemiluminescence Analysis by Molecular Emission Cavity Detection

Edited by

DAVID A. STILES

Department of Chemistry
Acadia University
Nova Scotia
Canada

A. C. CALOKERINOS

Department of Chemistry
University of Athens
Greece

ALAN TOWNSHEND

School of Chemistry
The University
Hull
UK

JOHN WILEY & SONS
Chichester / New York / Brisbane / Toronto / Singapore

Copyright © 1994 by John Wiley & Sons Ltd,
   Baffins Lane, Chichester,
   West Sussex PO19 1UD, England
   Telephone: National Chichester (0243) 779777
         International +44 243 779777

All rights reserved.

No part of this book may be reproduced by any means,
or transmitted, or translated into a machine language
without the written permission of the publisher.

Other Wiley Editorial Offices

John Wiley & Sons, Inc., 605 Third Avenue,
New York, NY 10158-0012, USA

Jacaranda Wiley Ltd, G.P.O. Box 859, Brisbane,
Queensland 4001, Australia

John Wiley & Sons (Canada) Ltd, 22 Worcester Road,
Rexdale, Ontario M9W 1L1, Canada

John Wiley & Sons (SEA) Pte Ltd, 37 Jalan Pemimpin #05-04,
Block B, Union Industrial Builidng, Singapore 2057

Library of Congress Cataloging-in-Publication Data
Flame chemiluminescence analysis by molecular emission cavity
   detection   edited by David Stiles.
      p.   cm. — (Chemical analysis) ; v. 129)
   Includes bibliographical references and index.
   ISBN 0-471-94340-1
   1. Molecular emission cavity analysis   I. Stiles, David, 1938–
   II. Series.
   QD79.P4F57   1994
   543′.0852 — dc20   93-33679
                                                CIP

British Library Cataloguing in Publication Data

A catalogue record for this book is available from the British Library

ISBN 0 471 94340 1

Origination at Alden Multimedia, Northampton, England
Printed and bound in Great Britain by Biddles ltd, Guildford, Surrey

CONTENTS

CONTRIBUTORS		ix
PREFACE		xi
ACKNOWLEDGEMENTS		xiii
CHAPTER 1	**THE HISTORY OF MOLECULAR EMISSION CAVITY ANALYSIS** *A. Townshend*	1
CHAPTER 2	**INTRODUCTION AND BASIC PRINCIPLES** *A. C. Calokerinos*	5

2.1	Flame Chemiluminescence	5
2.2	Historical Origins	6
2.3	Hydrogen Diffusion Flame	8
2.4	Salet Phenomenon	10
2.5	Molecular Emission Spectroscopy	12
2.6	Flame Photometric Detector	13
2.7	Molecular Emission Cavity Analysis	13
	2.7.1 Basic Principles	14
	2.7.2 The Importance of the Salet Phenomenon in MECA	25
	2.7.3 Sample Introduction into the Cavity	28
	2.7.4 Molecular Emissions and Spectra	35
2.8	Conclusions	38
References		39

CHAPTER 3	**INSTRUMENTATION AND AUTOMATION** *N. Grekas*	43

3.1	Introduction	43
3.2	Instrumentation	44
	3.2.1 Emission Burner Unit	44
	3.2.2 Cavity Probe and Holder Unit	45
	3.2.2.1 Cavity Probe	45
	3.2.2.2 Cavity Holder	50
	3.2.3 Optical Unit – Readout System	53
	3.2.4 Gas Generation Systems	54
3.3	Commercial Instruments	55
	3.3.1 Conventional MECA	55

		3.3.1.1	MECA-22 Spectrometer	55
	3.3.2	Gas Generation Detection		56
		3.3.2.1	MECA–VAP	56
			3.3.2.1.1 Design and Characteristics of MECA–VAP	57
			3.3.2.1.2 Design and Characteristics of the MEP–101/DIVAP–201	58
	3.3.3	Gas and High Performance Liquid Chromatographic Detection		59
3.4	**Automation**			61
	3.4.1	Conventional Automated Analysers		63
	3.4.2	Automated Gas Generation Analysers		65
3.5.	**Conclusions**			69
References				69

CHAPTER 4 SULPHUR, SELENIUM, AND TELLURIUM 71
E. Henden

4.1	**Introduction**			71
4.2	**Sulphur Compounds**			72
	4.2.1	Determination of Sulphur by Conventional MECA		73
		4.2.1.1	Inorganic Sulphur Compounds	73
		4.2.1.2	Organic Sulphur Compounds	76
		4.2.1.3	Sulphur Compounds in Detergents	79
		4.2.1.4	Sulphur in Solids	80
		4.2.1.5	Indirect Determination Based on the S_2 Emission	83
		4.2.1.6	Automated Conventional MECA	85
	4.2.2	Gas Generation Systems		85
		4.2.2.1	Determination of Sulphur	85
		4.2.2.2	Automated Gas Generation Systems	87
	4.2.3	Determination of Sulphur Compounds after Gas and Liquid Chromatographic Separation		89
4.3	**Selenium and Tellurium Compounds**			91
	4.3.1	Determination of Selenium and Tellurium by Conventional MECA		92
		4.3.1.1	Inorganic Selenium and Tellurium Compounds	92
		4.3.1.2	Organic Selenium and Tellurium Compounds	94
	4.3.2	Determination of Selenium and Tellurium by Gas Generation Systems		94
4.4	**Conclusions**			96
References				96

CHAPTER 5 ARSENIC, ANTIMONY, BORON, SILICON, GERMANIUM AND TIN 99
M. Burguera and J. L. Burguera

5.1	**Introduction**	99
5.2	**Arsenic and Antimony**	100
5.3	**Boron**	110
5.4	**Silicon**	114

5.5	Germanium	123
5.6	Tin	125
5.7	Conclusions	128
References		129

CHAPTER 6 NITROGEN, PHOSPHORUS, AND CARBON — 131
D. A. Stiles and A. Townshend

6.1	Introduction		131
6.2	Nitrogen Compounds		132
	6.2.1	General	132
	6.2.2	Indirect Methods	133
	6.2.3	Direct Methods	135
		6.2.3.1 Determination of Ammonia and Ammonium Ions	137
		6.2.3.2 Determination of Nitrite and Nitrate	143
6.3	Phosphorus Compounds		148
	6.3.1	General	148
	6.3.2	Inorganic Phosphorus Determinations	148
	6.3.3	Organic Phosphorus Determinations	154
6.4	Carbon Compounds		162
References			168

CHAPTER 7 HALOGENS AND METALS — 171
D. A. Stiles

7.1	Introduction		171
7.2	Halogens		171
	7.2.1	Fluorine	172
	7.2.2	Chlorine, Bromine and Iodine	174
7.3	Metals		187
7.4	Conclusions		192
References			192

INDEX — 195

CONTRIBUTORS

J. L. Burguera, University of Los Andes, Merida, Venezuela

M. Burguera, University of Los Andes, Merida, Venezuela

A. C. Calokerinos, University of Athens, Athens, Greece

N. Grekas, Research and Development Laboratory, Farmalex SA, Athens, Greece

E. Henden, Ege University, Izmir, Turkey

D. A. Stiles, Acadia University, Nova Scotia, Canada

A. Townshend, The University, Hull, UK

PREFACE

The 1960s were a period of rapid development for analytical atomic spectrometry. Atomic absorption spectrophotometry was developing as *the* technique for trace metal analysis. The first publications were appearing (Greenfield, Fassel) describing inductively coupled plasma (ICP) and its applications in multi-element analysis. Atomic fluorescence spectrometry (Winefordner) was announced, a technique that promised extreme sensitivity for metal and metalloid determination. And developments in the design of a practical furnace to replace the flame (Massmann) enabled electrothermal atomic absorption spectrophotometry to become a realistic analytical technique of great sensitivity.

The next decade seemed likely to be one of consolidation, where the new techniques would be refined, commercial equipment improved, and the range of applications increased. There remained, however, several problem areas. A serious deficiency of many of the atomic spectrometric techniques (with the exception of ICP atomic emission spectrometry) was their very poor sensitivity for some non-metals, especially sulphur, phosphorus and the halides. Yet sensitive flame emissions attributable to these elements were known, utilising the formation of excited state molecules in cool, hydrogen-based flames. The sulphur/phosphorus gas chromatographic detector was based on these emissions. Ronald Belcher's Analytical Group at the University of Birmingham, therefore, undertook an extensive programme of research into the development of cool flame techniques. The outcome was Molecular Emission Cavity Analysis (MECA), the subject of this monograph. It showed remarkable sensitivity, especially for the above elements, but it was also capable of detecting most non-metals and metalloids. It could accommodate solid, liquid and gaseous samples, and was able to analyse mixtures of species containing a common element (e.g. S^{2-}, SO_3^{2-} and SO_4^{2-}).

This book describes the development of MECA, its chemical basis, the instrumentation, and the various groups of elements which can be determined. Research continues in numerous institutions around the world, and it is hoped that this book will encourage them, and others, in their endeavours.

D. A. S.
A. C. C.
A. T.

ACKNOWLEDGEMENTS

We should like to thank all those whose time and effort have made possible the publication of this volume. However, we should like to pay special tribute to S. L. Bogdanski for the major contribution he made to the development of Molecular Emission Cavity Analysis during his time at the University of Birmingham.

One of us (D. A. S.) wishes to acknowledge the considerable support he has received from the Department of Environmental Science, University of Plymouth, where he has spent the past year on sabbatical leave.

June, 1993

D. A. Stiles
A. C. Calokerinos
A. Townshend

CHAPTER

1

THE HISTORY OF MOLECULAR EMISSION CAVITY ANALYSIS

ALAN TOWNSHEND

The University
Hull, UK

Molecular emission cavity analysis (MECA) was discovered by Belcher, Bogdanski, and Townshend (1–3) working in the Chemistry Department of the University of Birmingham more than twenty years ago. The events leading up to its discovery are a typical mixture of scientific curiosity, acute observation, serendipity and good research practice. They begin with (the late) Professor Belcher's intense interest in chemical reactions that might be useful in analytical chemistry. During bouts of insomnia he would resort to Feigl's *Spot Tests in Inorganic Analysis* (4), which is crammed with interesting reaction chemistry. At this time, 'flame tests' were of interest to Belcher, and we discussed more than once the test for tin whereby on treatment with zinc and hydrochloric acid, a solution containing tin generates gases (including hydrogen) which on introduction into the non-luminous flame of a Bunsen burner gives a characteristic blue colour (5). At the time, the nature of the blue colour was not revealed, but as is discussed in Chapter 6, it arises from emissions from SnO and (possibly) SnCl.

Intrigued by this test for tin and recalling his past experience with blow pipe analysis, in which numerous unusual flame colours are produced (6), Belcher sought and found further unusual flame tests in Feigl. The most interesting (7) was the test for bismuth 'which gives a cornflower-blue coloration to the hydrogen flame ... when the sample, mixed with alkaline earth carbonates, is placed in the flame. Among the common metals, only antimony and manganese compounds give analogous flame reactions, the former green-blue, the latter yellow' (8). Again, the nature of these emissions was not reported, but it was clear that they were not atomic emissions. As was later revealed, they were examples of **candoluminescence**, in which the emission is a luminescence emitted by solid (in this case alkali metal oxide) particles in the hydrogen flame. Indeed, it is further reported in Feigl (7) that 'all the rare earth oxides or carbonates when mixed with calcium carbonate, or precipitated together

---

Flame Chemiluminescence Analysis by Molecular Emission Cavity Detection
Edited by D. A. Stiles, A. C. Calokerinos and A. Townshend © 1994 John Wiley & Sons Ltd

with $CaCO_3$ ... likewise exhibit a luminescence' (9), which hints further of the need for the solid phase. These emissions were particularly interesting because they had detection limits in the nanogram region.

The test for bismuth was reported to detect down to 4 ng of bismuth, but no attempts had been made to base a quantitative analytical technique on the blue emission. On the arrival of a new PhD student (S. L. Bogdanski), therefore, it was decided to investigate the possibility of developing a quantitative flame technique based on candoluminescence. As the emission was emitted from a solid (e.g. CaO) held in the hydrogen-based flame, various devices were constructed in which the solid surface could be held in a reproducible position in the flame. Of these, the most successful, and probably the simplest, was a commonplace steel screw with a hexagonal aperture (max. diagonal 5 mm) in its head, which could be screwed into a bar and so positioned reproducibly in the flame. The aperture in the head of the screw was inlaid with a paste of the matrix (usually a $CaSO_4 \cdot CaO$ mixture) which, after being allowed to dry, was cut down to give a flat surface continuous with the surface of the screw head. A drop of analyte (e.g. Bi) solution was placed on the flat surface, and absorbed, after which the matrix surface was introduced into the flame, and the maximum emission intensity measured (10). This simple technique was successful, and procedures were subsequently developed for the determination of bismuth (11), manganese (12), praesodymium (13), terbium (13), europium (14) and cerium (14, 15), and other elements (16).

It was Bogdanski's early experiments with quantitative candoluminescence that led to the discovery of MECA. In order to re-use the matrix-holding screws for candoluminescence studies, most of the matrix was scraped out, and any residue dissolved in dilute phosphoric acid/sulphuric acid. After washing with distilled water, the screw head was introduced into the hydrogen flame for final cleaning, with interesting effect. As observed by Bogdanski (2) 'a blue emission was observed coming from the cavity space a few seconds after introduction of the screw into the flame. After reaching a maximum intensity, the blue emission subsided and was followed by a green emission which also reached a peak intensity and disappeared as the screw became incandescent'. These emissions, the blue S_2 emission from the H_2SO_4, and the green emission from the H_3PO_4, were intriguing, because they arose from what must have been extremely small residual amounts of the acids, and they were restricted to the interior of the cavity in the screw head. It seemed likely, therefore, that a very small volume of sample (containing sulphur or phosphorus compounds) could be introduced into the cavity, the cavity then inserted into the flame, and the contents thus analysed on the basis of these molecular emissions.

This was the birth of MECA (1, 2). Initial funding for the subsequent research was obtained from the Atomic Weapons Research Establishment, Aldermaston, UK. Results were so promising that a commercial MECA spectrometer was produced by Anacon Inc., Ashland, MA, USA (later Houston TX). A particularly important innovation was the design of a rotatable sample probe holder, which enabled sample to be introduced into a cavity held vertically upwards, out of the flame, and then the cavity could be rotated into a predetermined position in the flame. The technique was patented worldwide (3), and was announced at the first Euroanalysis symposium, held in Heidelberg, 1972.

The technique grew to provide an extremely sensitive, flame emission technique for the determination of a wide range of non-metals, metalloids and even metals, in solid, liquid or gaseous samples. It allowed speciation analysis (e.g. S^{2-}, SO_3^{2-}, SO_4^{2-}) in a single run (17), and has been adapted for gas (18) and liquid chromatographic (19) detection. From the Birmingham Analytical School and, latterly, Hull, 19 PhD students and numerous visiting researchers devoted their efforts to MECA. A total of 63 published papers originated from the Birmingham group, and numerous reviews (20–31) plotted the development of the technique. All these aspects are fully documented in the subsequent chapters in this book.

Research into cool-flame molecular emission spectroscopy including MECA, continues at numerous locations around the world. Some of the most prominent practitioners are contributors to this book. The extremely high sensitivity, its applicability to many elements (and species) of great environmental interest, the continuing development of more sensitive, or simpler detectors, and of array detection systems, should ensure that this will continue to be a very fruitful research area, and will provide elegant solutions to many analytical problems of current interest.

REFERENCES

1. R. Belcher, S. L. Bogdanski, and A. Townshend, *Anal. Chim. Acta*, **67**, 1 (1973).
2. S. L. Bogdanski, PhD Thesis, University of Birmingham, 1973.
3. R. Belcher, S. L. Bogdanski, Z. M. Kassir, and A. Townshend, Brit. Pat. Appl. 39443/72; US Patent 3,871,768, (Aug. 20, 1973) and patents based thereon. Also R. Belcher, A. Townshend, and S. L. Bogdanski, US Patent 3,981,585 (Sept. 21, 1976).
4. F. Feigl, *Spot Tests in Inorganic Analysis*, 5th Edn, Elsevier, Amsterdam, 1958.
5. Ref. 4, p. 109.
6. e.g. T. H. Richter, *Plattner's Manual of Qualitative and Quantitative Analysis with the Blow Pipe*, Chatto and Windus, London, 1875.
7. Ref. 4, p. 79.
8. J. Donau, *Monatsh*, **34**, 949 (1913).
9. A. Neunhofer, *Z. Anal. Chem*, **132**, 91 (1951).
10. R. Belcher, S. L. Bogdanski, and A. Townshend, *Talanta*, **19**, 1049 (1972).
11. R. Belcher, K. P. Ranjitkar, and A. Townshend, *Analyst*, **100**, 415 (1975).
12. R. Belcher, S. Karpel, and A. Townshend, *Talanta*, **23**, 631 (1976).
13. R. Belcher, K. P. Ranjitkar, and A. Townshend, *Analyst*, **101**, 666 (1976).
14. R. Belcher, T. A. K. Nasser, and A. Townshend, *Analyst*, **102**, 382 (1977).

15. R. Belcher, T. A. K. Nasser, L. Polo Diez, and A. Townshend, *Analyst*, **102,** 391 (1977).
16. R. Belcher, T. A. K. Nasser, M. Shahidullah, and A Townshend, *American Lab*, **61** (Nov., 1973).
17. R. Belcher, S. L. Bogdanski, D. J. Knowles, and A. Townshend *Anal. Chim. Acta*, **77,** 53 (1975).
18. R. Belcher, S. L. Bogdanski, M. Burguera, E. Henden, and A. Townshend, *Anal. Chim. Acta*, **100,** 515 (1978).
19. M. J. Cope and A. Townshend, *Anal. Chim. Acta*, **134,** 93 (1982).
20. R. Belcher, S. L. Bogdanski, S. A. Ghonaim, and A. Townshend, *Anal. Lett.*, **17,** 133 (1974).
21. R. Belcher, S. L. Bogdanski, S. A. Ghonaim, and A. Townshend, *Prog. Anal. Chem.* (1974) (in Russian).
22. R. Belcher, S. L. Bogdanski, D. J. Knowles, and A. Townshend, *Revista Chim.*, 1, 26 (1975).
23. R. Belcher, S. L. Bogdanski, and A. Townshend, *Inf. Chim.* No. 137, Nov. 1974 (in French).
24. R. Belcher, S. L. Bogdanski, and A. Townshend, *Zhur. Anal. Khim.*, 386 (1977).
25. R. Belcher, S. L. Bogdanski, O. Osibanjo, and A. Townshend, *Nigerian Bull. Chem.*, 31 (1978).
26. S. L. Bogdanski and A. Townshend, *Wiss. Z. Karl-Marx Univ.*, **28,** 377 (1979).
27. S. L. Bogdanski, M. Burguera, and A. Townshend, *CRC Crit. Rev. Anal. Chem.*, **10,** 185 (1981).
28. E. Henden, N. Pourezza, and A. Townshend, *Prog. Anal. Atom. Spec.*, **2,** 355 (1979).
29. M. Burguera, J. L. Burguera, and A. Townshend, *Rev. Roum. Chim.*, **5,** 761.
30. A. C. Calokerinos and A. Townshend, *Prog. Anal. Atom. Spec.*, **5,** 63 (1982).
31. M. Burguera, J. L. Burguera, and A. Townshend, *Acta Cient. Venezolana.*, **35,** 165 (1984).

CHAPTER

2

INTRODUCTION AND BASIC PRINCIPLES

A. C. CALOKERINOS

University of Athens
Athens, Greece

2.1	Flame Chemiluminescence	5
2.2	Historical Origins	6
2.3	Hydrogen Diffusion Flame	8
2.4	Salet Phenomenon	10
2.5	Molecular Emission Spectroscopy	12
2.6	Flame Photometric Detector	13
2.7	Molecular Emission Cavity Analysis	13
	2.7.1 Basic Principles	14
	2.7.2 The Importance of the Salet Phenomenon in MECA	25
	2.7.3 Sample Introduction into the Cavity	28
	2.7.4 Molecular Emissions and Spectra	35
2.8	Conclusions	38
	References	39

2.1 FLAME CHEMILUMINESCENCE

Some chemical reactions are accompanied by emission of light and the phenomenon is called *chemiluminescence* (CL). During a chemiluminescent reaction, an excited intermediate or a product releases its energy by emission of radiation. The complicated mechanism can be simplified and schematically shown by two steps. The first step is a *chemi-excitation* process which involves formation of the excited particle $(C^*)^1$ either by a direct or indirect chemiluminogenic mechanism. In *direct chemiluminescence*, chemi-excitation is achieved by a simple reaction

$$A + B \longrightarrow C^* \tag{1}$$

and the second step involves de-excitation by emission of radiation ($h\nu$)

$$C^* \longrightarrow C + h\nu \tag{2}$$

---

[1]Throughout this book, an asterisk will indicate an excited particle with the ability to radiate during de-excitation.

Flame Chemiluminescence Analysis by Molecular Emission Cavity Detection
Edited by D. A. Stiles, A. C. Calokerinos and A. Townshend   © 1994 John Wiley & Sons Ltd

De-excitation may also occur by ways other than by emission of radiation (radiationless de-excitation). Indirect CL occurs by an *energy transfer* process

$$A + B + C \longrightarrow AB + C^* \qquad (3)$$

and de-excitation proceeds as in direct CL.

CL reactions occur in the gaseous, liquid and solid phases. Solution CL has been known since 1877 when Radziszewski studied the chemiluminescent properties of 2,4,5-triphenylimidazole (lophine) (1). The best known example of solution CL is the oxidation of luminol (5-amino-2,3-dihydro-1,4-phthalazinedione) which was first observed by Albrecht (2). If CL occurs in a living organism or *in vitro* by the action of enzymes, the phenomenon is called *bioluminescence* (BL). Analytical applications of CL and BL are being developed and reviewed continually (3, 4).

Chemiluminescence in the gaseous phase occurs from gaseous reactants. The CL reaction may take place at ambient or at higher temperatures. If at least one of the reactants is a flame particle, then *flame chemiluminescence* (flame CL) occurs.

A typical example of CL in the gaseous phase at ambient temperature is the reaction between nitrogen monoxide and ozone (5).

$$NO + O_3 \longrightarrow NO_2^* + O_2 \qquad (4)$$

$$NO_2^* \longrightarrow NO_2 + h\nu \ (600-875\,nm) \qquad (5)$$

This reaction is used for the determination of nitrogen monoxide in urban air. Nitrogen dioxide does not interfere (6).

In flame CL, the gaseous analyte reacts with one of the particles of the flame. Most of these reactions generate excited molecules, but radicals or atoms may also form. The technique is known as *molecular emission spectroscopy* (MES). The excited molecules are not stable in the hot environment of common flames so cool flames must be used if they are to be observed. The hydrogen flame is that most commonly used in cool flame CL and will be discussed in detail.

The main difference between flame CL and flame photometry is in the excitation process. In flame photometry, where thermal excitation occurs, atoms are the predominant emitting species and, in accordance with the Maxwell–Boltzmann law, the intensity increases with temperature. In flame CL, excited molecules or radicals which are not stable in high temperatures predominate. Nevertheless, the distinction between chemiluminescence and radiation by thermal excitation is not always clear and there are examples where both phenomena occur simultaneously.

2.2 HISTORICAL ORIGINS

Molecular emissions were observed by most early investigators, working on analytical applications of flames. Banded molecular emissions from sulphur, phosphorus, selenium, tellurium and many other elements have been described by Bunsen, Mitscherlich, Brewster and other early spectroscopists.

A detailed description of the blue colouration of a hydrogen flame by a sulphur compound was presented by Mulder in 1864 (7). The brightness of the emission and the appearance of well-separated identical spectral lines which were obtained after the radiation had been passed through a prism spectrometer attracted further examination.

In 1865, Barrett (8) performed a series of experiments in which a hydrogen flame was brought into contact with several different solids in his laboratory. In all cases, he noticed a blue colouration at the point where the flame was in contact with the solid surface. The emission was initially intense, but as the point of contact heated up, it disappeared, only to appear once again when a different spot on the same solid was heated. He soon realised that this emission was due mainly to sulphur from dust and decomposition products of coal burning, which had deposited on all solids in his laboratory. The decrease in emission intensity as the solid heated up was attributed to the consumption of the sulphur at the point of contact, and to the increasing temperature of the surface. Barrett realised that the emission was very sensitive and proposed it as the basis of a test for examining whether sulphur is present on the surface of solids exposed to atmospheric air. He also mentioned that only the hydrogen flame generated the sulphur emission and briefly discussed the emission of the flame itself.

In 1869, Salet (9,10) examined the blue colouration from sulphur within a hydrogen flame through a spectroscope and took appropriate precautions necessary to avoid the kind of surface contamination described by Barrett. The spectrum, showing emission lines in the green and blue regions, proved to be identical to that obtained by Mulder.

Salet noticed that the blue emission was more intense in the central core of the flame and did not appear at the periphery where the flame temperature was higher. He also observed that when the hydrogen flame impinged on the surface of sulphur-containing solids, the characteristic blue emission appeared. As the solid heated up, the emission intensity decreased and the point of contact between the flame and the surface had to be changed frequently to maintain the initial intensity. Barium sulphate showed no blue emission since it does not decompose at the low temperature provided by the flame. In a further experiment, Salet mixed barium sulphate with a phosphate salt and heated the mixture in the flame at the end of a platinum thread. He noticed that the sulphur emission appeared only when the flame impinged on the relatively cool platinum surface. The emission spectrum showed bands from sulphur and phosphorus, the latter appearing in the flame as a green halo. Although the author does not mention which phosphate compound was used, it seems that he was able to remove cationic interferences on the sulphur emission to some extent and to generate sulphur emission from the refractory barium sulphate.

After the work of Salet, sporadic reports and very few publications on molecular emissions from flames appeared. Analytical applications of flame CL appeared at the beginning of the 1960s. At that time, work on the design and applications of the flame photometric detector for gas chromatography (GC) and of sheathed flames for MES was initiated. The sensitivity of MES attracted many analytical chemists to pursue further work on the determination of sulphur, phosphorus and other non-metals and metalloids by the technique of aspiration into the hydrogen flame.

2.3 HYDROGEN DIFFUSION FLAME

Hydrogen is the fuel gas of the flame and is supplied to the burner along with oxygen or air (*premixed hydrogen flame*). Alternatively, an inert gas such as nitrogen can be used as a diluent. In this case, combustion is achieved by diffusion of atmospheric oxygen into the gas mixture and a *hydrogen diffusion flame* appears at the top of the burner.

The flame is ignited (11) by the reaction:

$$H_2 + O_2 \longrightarrow 2OH\cdot \qquad (6)$$

and is maintained by the following reaction:

$$H_2 + OH\cdot \longrightarrow H + H_2O \qquad (7)$$

This is the main reaction in the flame, but chain branching may also occur:

$$O_2 + H \longrightarrow OH\cdot + O \qquad (8)$$

$$H_2 + O \longrightarrow OH\cdot + H \qquad (9)$$

The maximum temperature of the hydrogen diffusion flame is about $1000°C$ and is much lower than that of the corresponding hydrogen–air premixed flame (12). Since mixing of hydrogen and atmospheric oxygen is controlled by diffusion, the amount of oxygen which penetrates the flame from its edges to the centre is drastically reduced. As a result of this difficulty in mixing oxygen with hydrogen, the temperature in the centre of the flame is lower than that at the edge. The temperature profile of the hydrogen diffusion flame is schematically shown in Figure 2.1. Temperature changes within the flame are accompanied by a variation in the nature and concentration of flame particles. The edge of the flame is rich in oxygen and also contains water vapour from the main chemical reaction of the flame, while the inner part of the flame is rich in atomic and unburned hydrogen (see Figure 2.1).

The central part of the flame is coolest and is usually responsible for CL reactions with hydrogen. These reactions occur by direct mechanisms, for example:

$$SnCl_2 + H \longrightarrow SnCl^* + HCl \qquad (10)$$

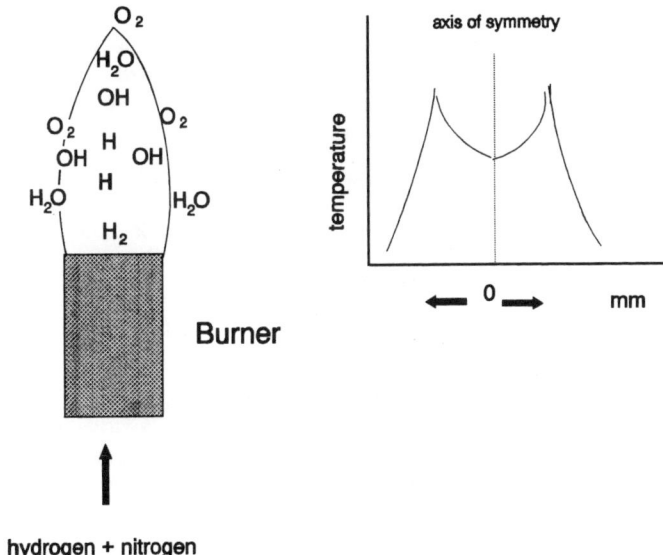

Figure 2.1. Schematic diagram of temperature and concentration profile of the cross section of a hydrogen diffusion flame.

or indirect mechanisms in which energy is transferred by or during the recombination reactions

$$Na + H + H \longrightarrow Na^* + H_2 \quad (11)$$

$$Na + H + OH\cdot \longrightarrow Na^* + H_2O \quad (12)$$

The radiation from chemically excited sodium atoms is identical to emission from thermal excitation. Reactions 11 and 12 are exothermic and provide 431 and 497 kJ mol^{-1}, respectively (13).

The ability of the hydrogen diffusion flame to act as a temperature and concentration gradient across its body can be shown with a simple experiment. If tin(II) bromide is aspirated into the flame, three coloured regions appear (Figure 2.2) (14). In region 1, the concentration of atomic hydrogen is low and the excited particles are formed by the reaction:

$$SnBr_2 + H \longrightarrow SnBr^* + HBr \quad (13)$$

This part of the flame appears green.

After de-excitation, SnBr flows to higher portions of the flame where the concentration of atomic hydrogen is higher than in region 1. Thus, region 2 appears

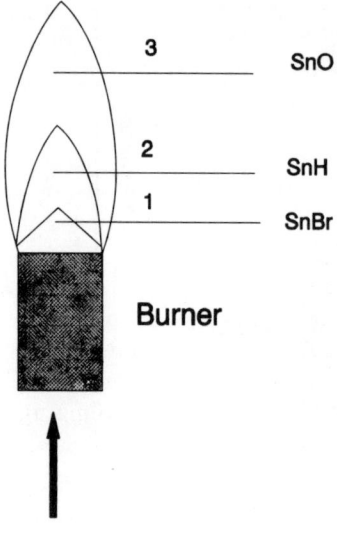

Figure 2.2. Schematic diagram of the distribution of predominating emitting species from the aspiration of tin(II) bromide in a hydrogen diffusion flame.

red due to the reaction:

$$SnBr + 2H \longrightarrow SnH^* + HBr \quad (14)$$

Unreacted SnBr from region 1 and SnH from region 2 flow upwards and sideways towards the periphery of the flame. This part of the flame is rich in molecular oxygen due to contact with atmospheric air and hydroxyl radicals from reactions 8 and 9. The following reactions occur:

$$SnH + OH\cdot \longrightarrow SnO^* + H_2 \quad (15)$$

$$SnH + O_2 \longrightarrow SnO^* + OH\cdot \quad (16)$$

$$SnBr + OH\cdot \longrightarrow SnO^* + HBr \quad (17)$$

and region 3 appears blue. The same sequence of reactions occur if tin(II) chloride is aspirated instead of tin(II) bromide, the only difference being that $SnCl^*$ is produced in region 1.

2.4 SALET PHENOMENON

During his studies on the sulphur emission generated within the hydrogen diffusion flame, Salet realised that the blue emission appears at moderate temperatures. Therefore, he decided to isolate the element and establish the optimum temperature for the emission. He introduced a thin tube of platinum alongside the flame and allowed water to flow slowly through the tube in order to maintain a temperature of about 100°C in that part of the flame (15). This was the minimum temperature which prevented condensation of water vapours from the flame on the outer surface of the tube. The tube was relatively cool compared to the flame and the sulphur emission was enhanced. A bright blue sheath appeared around the tube which was eventually covered by a thin skin of sulphur.

Salet noticed that the sulphur emission disappeared when the temperature increased. In order to estimate this temperature, he introduced a very thin platinum wire into the flame in such a position as to enable a comparison to be made between the emission intensity from sulphur and that from the hot metal. Initially, the metal served as a cooling agent but as the temperature increased, the blue emission around it decreased and eventually disappeared when the metal became red hot. He concluded that the emission disappears when the temperature exceeds 500°C.

Salet also observed that cooling can be achieved by impinging the flame on a sheet of ice placed on a thin strip of platinum (16) or by blowing a strong current of chilled air onto the flame (17,18). He also noticed that a cool flame environment is necessary for the green phosphorus emission.

In conclusion, Salet was the first to notice and study the enhancement of sulphur and phosphorus emissions from a hydrogen diffusion flame when a cool body is placed in its hot environment. The importance of this discovery was not realised until many years afterwards and is now known as the *Salet phenomenon*.

Later Studies on the Salet Phenomenon

No further investigation on the Salet phenomenon appeared in the literature until the development of molecular emission spectroscopy and of sulphur and phosphorus detectors for gas chromatography.

Crider (19) noticed that when air containing sulphur dioxide was introduced into a hydrogen flame, no emission was observed until the concentration became greater than 10 mg L^{-1}. However, when a borosilicate glass tube was used as a flame sheath, the blue emission appeared outside the combustion zone of the hydrogen–air flame and near the wall of the sheath, even with concentrations less that 1 mg L^{-1} of sulphur dioxide (Figure 2.3).

The use of sheathed flames offers high sensitivity because the walls of the sheath are maintained at a temperature lower than that of the flame. Thus, sulphur and phosphorus emissions are enhanced in accordance with the Salet phenomenon. Further work on sheathed flames accompanied the development of the flame photometric detector for gas chromatography.

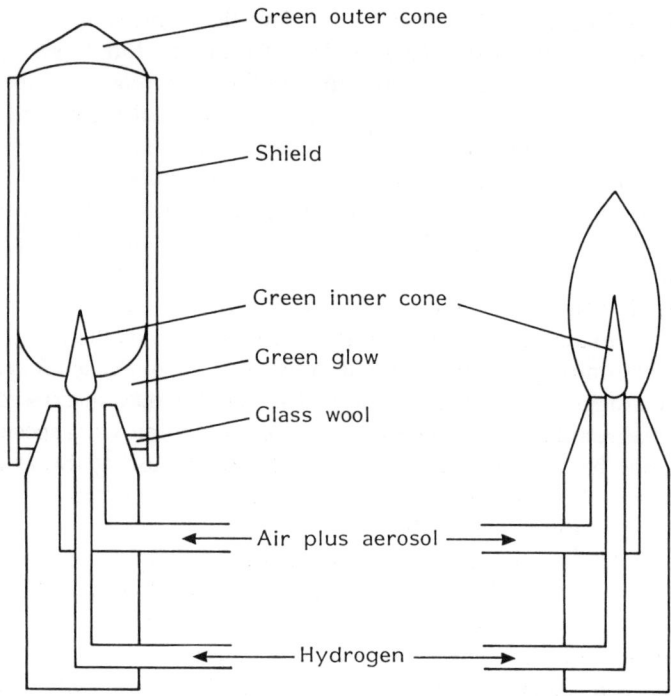

Figure 2.3. Schematic diagram of simple and sheathed hydrogen–air flame and coloured regions formed when a phosphorus compound is aspirated in both flames (22).

2.5 MOLECULAR EMISSION SPECTROSCOPY

Dagnall et al. (20) made a thorough investigation of the emissions from sulphur compounds aspirated into various flames. Hydrogen diffusion flames with nitrogen or argon as a diluent proved to be the most useful ones for analytical purposes. The emission spectrum recorded was attributed to the formation of excited S_2 molecules although it was not possible to obtain a response from all compounds introduced because the flame was not hot enough to decompose them. When a hydrogen–air flame was used, it promoted the breakdown of more sulphur compounds, but as expected, reduced sulphur emission intensities were observed because the hot environment of this flame destroyed the precursors.

This problem was overcome by using a quartz or borosilicate glass flame sheath. In this configuration, the sulphur-containing analyte was first broken down in the hottest part of the flame to form excited molecules. These then became stabilised on the cool inner surface of the sheath to generate the characteristic blue sulphur emission.

The same phenomenon was observed when a solution of orthophosphoric acid was nebulized into a hydrogen diffusion flame. In this case, a green emission attributed to the excited HPO species (21) increased in intensity when a sheathed flame was used (22).

When an inert gas sheath was used instead of a glass sheath, no enhancement of emission intensities from S_2 and HPO was observed, suggesting that a solid surface is vital for this enhancement (23). Further studies involving electron microprobe analysis of a Vycor tube used as a flame sheath for S_2 and HPO emissions showed that sulphur and phosphorus deposited on the surface (24) and seemed to confirm its role in the enhancement process. This same study reported that a chilled wall, introduced into the flame, enhanced emission intensities.

The analytical applications of sheathed flames in MES have been extensively investigated. Compounds which contain sulphur (25–30), phosphorus (31–33), boron (34), nitrogen (35) and the halogens (36, 37) have been determined in this fashion. The spectra and other characteristics of various emitting species generated by flame CL have also been studied (38–40), but a more thorough review is beyond the scope of this book.

2.6 FLAME PHOTOMETRIC DETECTOR

The flame photometric detector (FPD) for gas chromatography was first described by Brody and Chaney (41) and is used for the determination of volatile sulphur- and phosphorus-containing compounds. The principle of operation is the formation of excited S_2 and HPO molecules within a shielded hydrogen flame. Nitrogen is used as the carrier gas and oxygen is mixed at the outlet of the column (Figure 2.4). Oxygen and nitrogen are mixed in the same ratio as in air. Hydrogen is brought directly into the burner and the gas mixtures are burned in a hollow tip which shields the flame from direct view of the photomultiplier tube. This arrangement is necessary in order to prevent interfering emissions from the combustion reactions reaching the detector. When sulphur or phosphorus compounds are present, the characteristic emissions appear above the flame shield and the intensity is monitored by the photomultiplier. Selection of wavelength can be achieved by filters with maximum transmittance at 394 and 526 nm for sulphur and phosphorus, respectively. In a modified dual detection design, two filters have been used for the simultaneous measurement of both elements (42). A modified dual FPD has been proposed by Chester as a detector for HPLC (43). In this configuration, quenching of the molecular emissions is reduced by a continuous flow of organic solvent into the detector.

2.7 MOLECULAR EMISSION CAVITY ANALYSIS

Molecular Emission Cavity Analysis (MECA) is a flame chemiluminescence

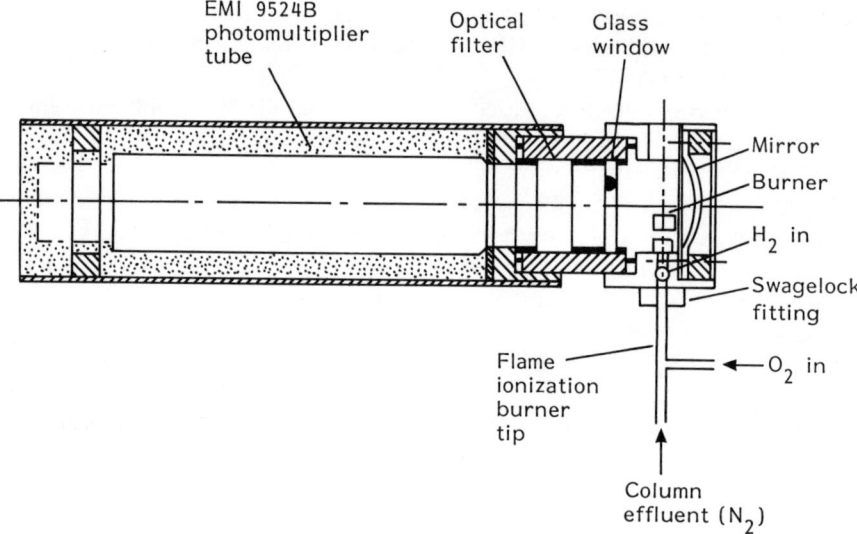

Figure 2.4. Schematic diagram of flame photometric detector for gas chromatography (41).

technique which is based on the generation of excited molecules, radicals, or atoms within a hydrogen diffusion flame. The excited species are formed by direct or indirect CL mechanisms and are confined within the inner space of a small cavity which is positioned at a preselected point in the flame environment. The cavity is responsible for the distinct characteristics of MECA over other emission techniques where the sample is aspirated into the flame. Hence, the cavity is the *cell for the chemiluminometric reactions* and promotes excitation due to chemiluminescent rather than thermal processes.

Since the initial observation (44) (see Chapter 1), a plethora of elements have been found to radiate characteristic emissions within the cavity (Table 2.1.). Nevertheless, the list of elements is not yet complete because research has been directed primarily into the area of non-metals and metalloids in order to propose MECA as a complementary technique to atomic emission, absorption and fluorescence techniques. Furthermore, since MECA is an emission technique it has the potential for simultaneous determination of two or more elements although this possibility has never been investigated.

2.7.1 Basic Principles

Flame

The flame normally used in MECA is the hydrogen diffusion flame described in

Table 2.1. Emissions generated within the MECA cavity and the most commonly used wavelengths for analytical measurement

Element	Emitting particle	Wavelength (nm)	Cavity type
Antimony	SbO–O continuum	355	oxy-cavity
Arsenic	AsO–O continuum	400	oxy-cavity
Boron	BO_2	518	oxy-cavity
Bromine	CuBr	494	copper
	InBr	375	indium-lined
Cadmium	Cd	326.1	normal
Carbon	CH	431.5	oxy-cavity
Chlorine	CuCl	532	copper
	InCl	360	indium-lined
Gallium	GaBr	350	normal
	GaI	391	normal
Germanium	GeCl	455	normal
Iodine	CuI	510	copper
	InI	410	indium-lined
Lithium	Li	670.8	normal
Nitrogen	NO–O continuum	500	oxy-cavity
Phosphorus	HPO	528	normal
Selenium	Se_2	411	normal
Silicon	SiO	540	oxy-cavity
Sulphur	S_2	384	normal
Tellurium	Te_2	500	normal
Thallium	Tl	377.5	oxy-cavity
Tin	SnO	485	normal

Section 2.3. This flame promotes CL reactions by direct (Reaction 10) or indirect mechanisms (Reactions 11–12). The reactions which occur in the flame (Reactions 6–9) generate various species but only the hydroxyl radical emits radiation. Thus, the flame spectrum shows the OH band at 306 nm and a very weak continuum band covering the visible region. The OH emission intensity from the hydrogen diffusion flame is about 40 times weaker than in a premixed hydrogen–air flame maintained on the same burner head. The feeble flame emission allows weak molecular emissions from added species to be detected, provided that stray light is excluded from the detector.

Aspiration of the analyte into the hydrogen diffusion flame for generation of CL has some drawbacks:

1. The low temperature of the flame may lead to formation of solid particulates which reduce the production of emitting species, and therefore, limits sensitivity.

2. The solvent and other components of the analyte solution may alter the temperature and disturb the radical distribution and concentration within the

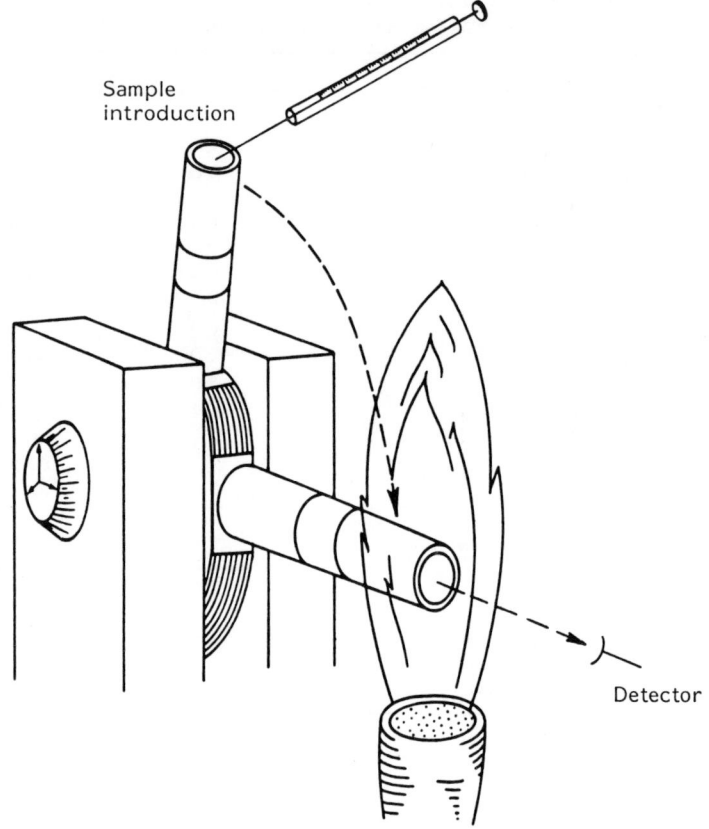

Figure 2.5. Conventional MECA cavity introduction into flame by manual rotation (82).

flame body. As a result, the intensity of the desired emission and, perhaps, the background emission as well, will be changed during the measurement.

3. The distribution of the sample vapours within the flame may be responsible for the generation of more than one emitting species, such as with tin(II) bromide (Reactions 13–17). Production of excited species will also be affected by the difference in temperature at various points in the flame. Moreover, the emitting species will spread out and their intensity per unit area of flame facing the detector will be low.

4. The residence time of the analyte within the flame is short and cannot be increased, since it is controlled mainly by the flow rate of the support gas.

The problems associated with aspirating sample solutions into a hydrogen diffusion flame for chemiluminogenic reactions are minimized by using a MECA cavity for sample introduction.

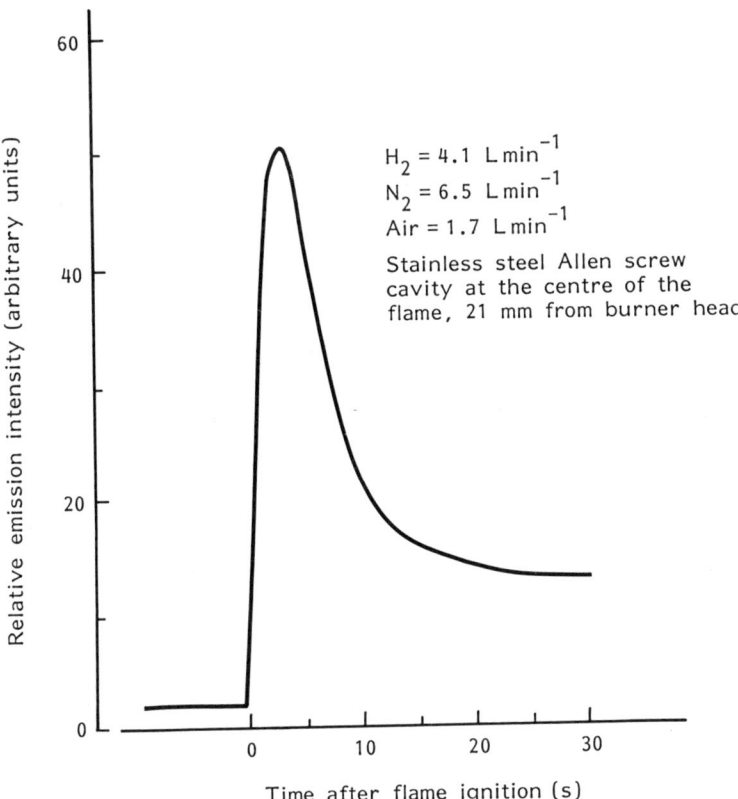

Figure 2.6. S_2 emission profile at 384 nm from 2 μL of 10 μg sulphur mL^{-1} as aqueous thiourea injected in a stainless steel MECA cavity (49).

Cavity

Extensive basic and applied research on MECA as an analytical technique has led to the development of two different ways of using the cavity as a means for sample introduction into the flame environment. Those ways define two different techniques, namely *conventional MECA* and *gas generation MECA* detection.

In conventional MECA, a few microlitres of solution or several milligrams of solid are deposited in the cavity, which typically is 8 mm in diameter and 5 mm in depth. The cavity is then introduced rapidly and reproducibly into the flame (Figure 2.5). The cavity temperature then increases from ambient to its maximum value, which depends on the material from which the cavity is made and conditions within the flame.

During this heating period, a series of physical and chemical changes occur, which are accompanied by generation of the characteristic molecular emission.

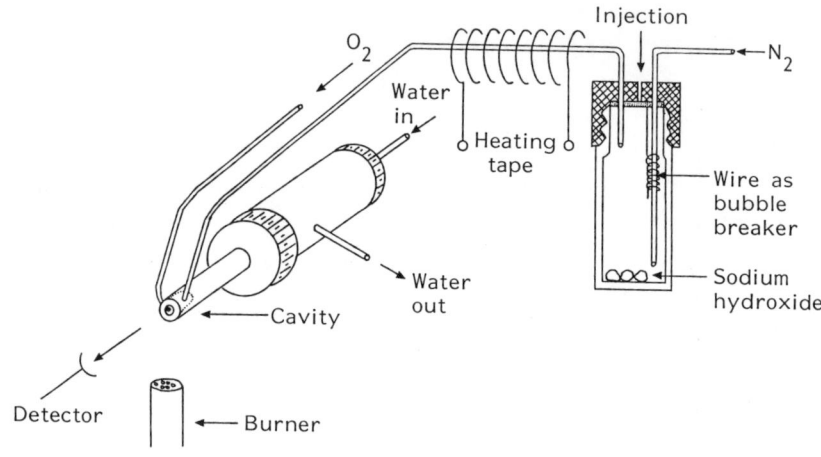

Figure 2.7. Schematic diagram of a gas generation MECA system for the determination of ammonium. Reproduced by permission of The Royal Society of Chemistry from Belcher et al. (66).

A typical example of a recorded response from thiourea, which generates the S_2 emission, is shown in Figure 2.6. The cavity remains in the flame as long as required for the emission to reach a maximum value and then decreases to zero. It is then removed from the flame and cooled before the process is repeated.

In gas generation MECA detection, the analyte is usually converted into a vapour within a closed reactor which is continuously purged by a carrier gas, such as nitrogen or argon. The analyte vapour is then swept through a side duct (Figure 2.7) into the cavity where the MECA emission is generated and recorded. Other MECA configurations also exist and will be described in later parts of this book.

Functions of the Cavity

The processes which may occur inside the cavity during conventional MECA are:

1. Solvent evaporation, accomplished by blowing hot air onto the cavity or by heating the cavity on a hot plate before it is transferred to the cavity holder.

2. Analyte evaporation and decomposition, in which the analyte undergoes one or more of the following processes:

(a) *Boiling*, e.g. sulphuric acid (45).

(b) *Sublimation*, e.g. iodine and selenium dioxide. Iodine sublimes directly in a heated cavity while selenium dioxide generates the Se_2 emission at a cavity temperature of about 315°C, the temperature at which the dioxide sublimes (46).

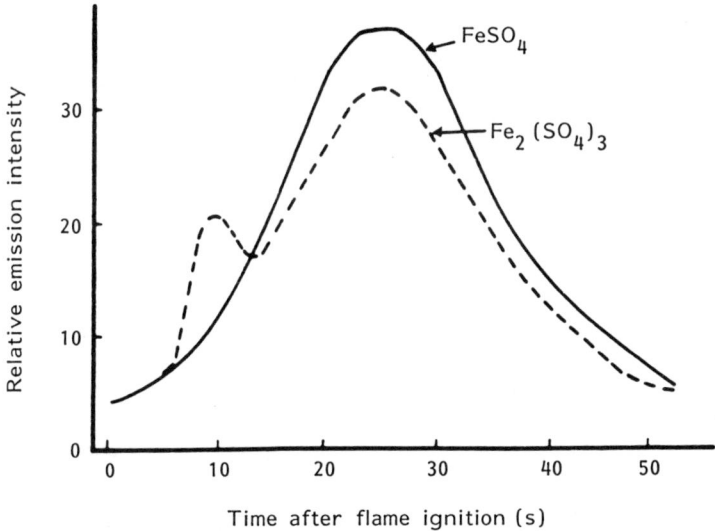

Figure 2.8. S_2 emission profiles from iron(II) and iron(III) sulphates (51).

(c) *Thermal decomposition*, e.g. iron(II) sulphate, which decomposes as follows:

$$FeSO_4 \longrightarrow FeO + SO_3 \qquad (18)$$

The sulphur trioxide produced in this reaction is responsible for generation of the S_2 molecular emission.

(d) *Reductive breakdown*, e.g. iron(III) sulphate, which decomposes in the reductive atmosphere of a hydrogen diffusion flame as follows:

$$Fe_2(SO_4)_3 + 2H \text{ (flame)} \longrightarrow 2FeSO_4 + H_2SO_4 \qquad (19)$$

Hence, when sulphuric acid is heated in a stainless-steel MECA cavity, it first boils generating the S_2 emission. This is subsequently followed by a second response due to the thermal decomposition of iron(II) sulphate formed by reaction of the sulphuric acid with the cavity material (Figure 2.8). When a carbon cavity is used, the first response appears even faster because reductive breakdown which aids in the formation of S_2 is easier.

Other examples of compounds which generate characteristic emissions before the cavity reaches a temperature at which they thermally decompose are the sulphates of copper(II), silver(I), cadmium(II) and mercury(II). With a stainless-steel cavity, the order in which emissions occur is related to the redox potential of the metal ion. Again, when the stainless-steel cavity is replaced by one of carbon coated with pyrolytic carbon, the emissions appear earlier because reductive breakdown enhances the decomposition process.

For copper(II) sulphate, the general reaction which occurs in a stainless-steel cavity is:

$$CuSO_4 + 2H \text{ (flame)} \longrightarrow Cu + SO_3 + H_2O \qquad (20)$$

and in a carbon cavity is:

$$CuSO_4 + C \text{ (cavity)} \longrightarrow Cu + SO_3 + CO \qquad (21)$$

Both reactions are followed by the chemiluminogenic sequence

$$SO_3 \longrightarrow S_2^* \longrightarrow S_2 + h\nu \qquad (22)$$

(e) *Catalytic breakdown*, e.g. thiodiphenylamine and its derivatives, promethazine, chlorpromazine, and perphenazine which generate S_2 at a cavity temperature of about 100°C, well below their normal decomposition temperature (47). The emission occurs through the following reaction:

[Structural reaction: thiodiphenylamine derivative + 2H $\xrightarrow{\Delta, \text{metal}}$ carbazole derivative + H_2S]

The reaction is probably catalysed by the metallic constituents of the cavity, although the exact mechanism is not yet understood.

3. Generation of excited species. During this step, direct or indirect chemiluminescent reactions, which produce excited species, occur. A typical example is the generation of the excited diatomic sulphur molecule (48)

$$\text{Sulphur compound} \xrightarrow{\text{reduction}} H_2S \qquad (23)$$

$$H_2S + H \longrightarrow HS + H_2 \qquad (24)$$

$$HS + H \longrightarrow S + H_2 \qquad (25)$$

$$S + S \longrightarrow S_2^* \qquad (26)$$

$$S_2^* \longrightarrow S_2 + h\nu \; (300\text{--}460 \text{ nm}) \qquad (27)$$

The energy for excitation of the diatomic sulphur molecule is provided by the recombination reactions 11 and 12.

4. De-excitation and emission (a) *Conventional MECA*. In conventional MECA, the vaporization process (step 2) determines *the time at which an emission reaches maximum intensity after the cavity has been introduced into the flame*. This is abbreviated as t_m.

The t_m value is a characteristic parameter for each compound but is also affected by the temperature of the cavity. This effect is expected, since a higher cavity temperature will accelerate vaporization and decomposition processes.

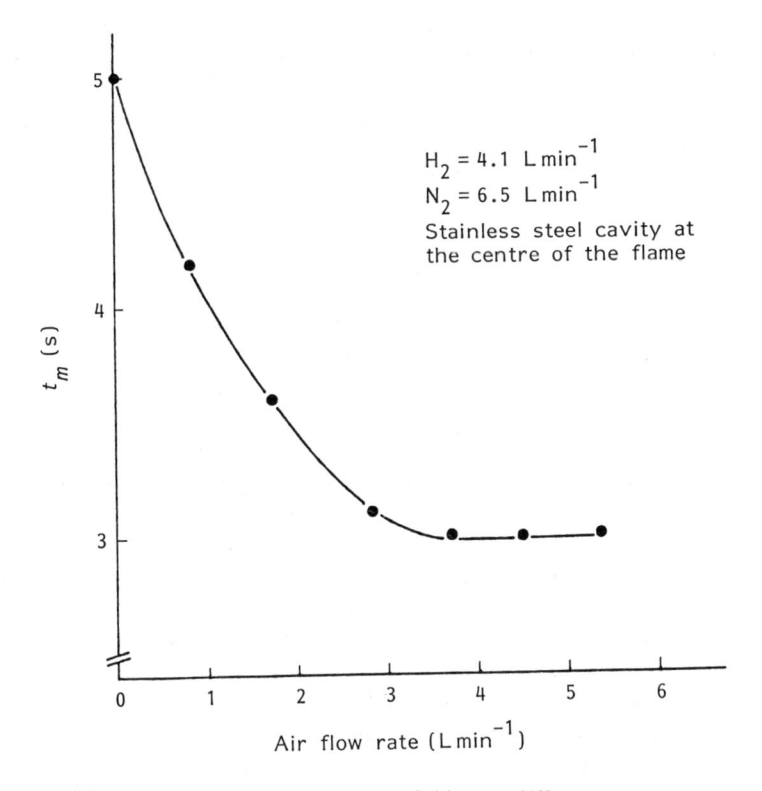

Figure 2.9. Effect of air flow on the t_m value of thiourea (49).

Hence, the t_m value of thiourea (Figure 2.9) decreases when air is added to a hydrogen diffusion flame because its temperature is increased (49).

The t_m value is an experimental parameter which can be used for speciation (Table 2.2). For instance, a separation of sulphur emissions would allow the successful measurement of anions in acid rain.

(b) *Gas generation MECA detection.* In gas generation MECA detection, the cavity is water-cooled to allow it to remain permanently within the flame (Figure 2.10). Since the cavity is cooled externally[1], its temperature remains constant and the only processes that can occur are formation, excitation and emission. Therefore, the analyte must be supplied to the cavity as a gas or vapour after a chemical or physical transformation, and the t_m concept is not valid. Typical reactions used in gas generation MECA detection are summarized in Table 2.3.

---

[1]The minimum temperature for cavity cooling is just over 100°C to avoid condensation of water vapours from the flame reaction 7.

Table 2.2. MECA t_m Values for Some Sulphur Anions (82)

Anion	S^{2-}	$S_2O_5^{2-}$	SCN^-	SO_3^{2-}	$S_2O_3^{2-}$	SO_4^{2-}	$S_2O_8^{2-}$
t_m, s	1.4	1.7	2.4	2.4	3.0	8	8

Conditions: 4.5 L H_2 min^{-1}, 6.5 L N_2 min^{-1}; sulphide in 0.01 mol L^{-1} ammonium phosphate buffer pH 7 and all other anions in 0.1 mol L^{-1} H_3PO_4.

Position of Cavity within the Flame

As already stated, the hydrogen diffusion flame exhibits temperature and concentration gradients along its length (Figure 2.1). A particular compound or group of compounds require a given concentration of hydrogen to generate maximum emission intensity and, in the case of conventional MECA, they also require a given temperature to vaporize. Since the sample is introduced into the flame via the cavity, its position within the hydrogen concentration gradient and, if required, within the temperature gradient is very critical and must be selected by optimizing for sensitivity and reproducibility of emission.

The advantage of using a cavity rather than the aspiration method for introducing sample into the flame is best illustrated with tin(II) chloride. Aspiration of tin(II) chloride into the flame generates a variety of emitting species, similar to those shown in reactions 13–17 for tin(II) bromide. If tin(II) chloride is introduced into the flame with conventional MECA, the characteristic SnCl emission appears within the cavity indicating an environment of low temperature, low concentration of hydroxyl radicals, and a modest concentration of hydrogen atoms. In this case, one specific molecular emission has been identified, and all conditions are then optimized for maximum radiation intensity at this one wavelength.

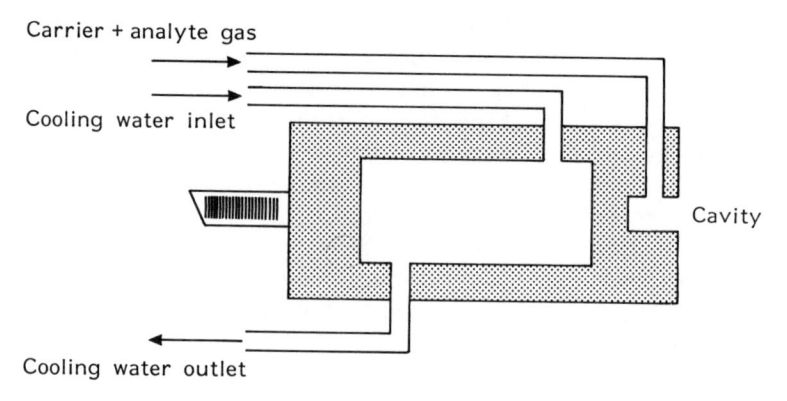

Figure 2.10. Schematic diagram of water-cooled MECA cavity for continuous operation within the flame.

Table 2.3. Gas generating reactions used for MECA detection

Analyte	Reactant	Generated gas
B	$CH_2OH + H_2SO_4$	$(CH_3)_3BO_3$
As	$NaBH_4$	AsH_3
Sb	$NaBH_4$	SbH_3
Sn	$NaBH_4$	SnH_4
Se	$NaBH_4$	H_2Se
Te	$NaBH_4$	H_2Te
Si	$F^- + H_2SO_4$	SiF_4
F^-	$SiO_4^{4-} + H_2SO_4$	SiF_4
S^{2-}	H^+	H_2S
SO_3^{2-}	H^+	SO_2
NH_4^+	OH^-	NH_3
NO_2^-	$H^+ + I^-$	NO

The fraction of flame atoms, radicals and molecules which enter the inner confines of the cavity is not only governed by the position of the cavity within the flame but also by the cavity angle. This determines the degree of contact of flame gases with the injected analyte. For small cavities optimum values lie in the range 5–20° below horizontal. If the angle is greater than 20°, the reproducibility of response is reduced, while decreased emission intensities are obtained through inefficient mixing of analyte with flame gases when this angle is less than 5°. The optimum angle is 7° for cavities of 8 mm diameter and 5 mm depth. Generally, as the cavity size increases, the angle becomes less critical since the large opening of the aperture allows adequate mixing of gases to occur.

Cavity Design and Material of Construction

Generally, the cavity size must be very small compared to that of the flame, to prevent the possibility of flame gas turbulence which will alter the degree of background emissions. Background radiation can be minimized if an iris diaphragm is placed in front of the photomultiplier tube of the detector assembly. However, the small size of the cavity does make its position within the flame very critical for generation and measurement of molecular emissions.

Theoretically, for any given analytical MECA procedure, an optimum cavity size should be determined so that maximum and reproducible emission intensities are obtained. In the case of an aluminium cavity, used to measure S_2 emission intensities, optimum dimensions were shown to be 5 mm in depth and 8 mm in diameter (50).

Reflectivity

During emission, some radiation is reflected from the back wall of the cavity and, hence, small changes in the surface reflectivity will alter emission intensities. Reflectivity depends on the material of construction, temperature, and wavelength of emission, but its effect on intensity measurements can be easily understood.

(a) *Conventional MECA.* When a stainless-steel cavity is alternately heated for brief periods in a flame, and then cooled, its inner surface changes from straw-yellow to blue and then to deep violet due to the formation of an oxide layer of varying thickness. Bogdanski (51) noticed that the S_2 emission from volatile sulphur compounds which do not leave a deposit on the surface varies with the degree of reflectivity of the cavity surface and can be reduced by as much as 50% in a cavity where the surface is highly oxidized. Therefore, in conventional MECA, reproducible results can only be obtained if the reflectivity of the cavity surface does not vary. This can be achieved by maintaining a constant residence time of the cavity within the flame and by ensuring that the cooling time prior to sample injection is also reproducible. A severe reduction in reflectivity occurs when the analyte leaves solid deposits inside the cavity. These effects occur only with metal cavities. Carbon and silica lined cavities are not subject to this interference.

(b) *Gas generation MECA detection.* The reflectivity effect in gas generation MECA detection is less severe than in conventional MECA due to the lower temperature of cavity operation. However, in the determination of ammonia using an aluminium cavity (52), a gradual decrease in background emission was noticed during the determination. The effect was only noticed when the photometric detector was operated at the highest sensitivity and was attributed to oxide formation on the inner surface of the cavity. When the inside surface of the cavity was polished with fine emery paper, the background emission returned to its original value. The reflectivity effect can be partially overcome by analysing standard solutions frequently, but might expect to be eliminated entirely by using a cavity coated internally with gold leaf.

Incandescence

Cavities which heat up continuously after introduction into the flame emit incandescent radiation. According to black body radiation theory, the degree of incandescence increases with wavelength and the emissivity of incandescing material (Figure 2.11). Thus, emissions at short wavelengths, such as that of S_2 at 384 nm, are virtually unaffected by incandescent background emissions from the cavity.

Incandescent background emission is a serious problem when the analyte vaporizes at elevated temperatures and emits at long wavelengths. The effect

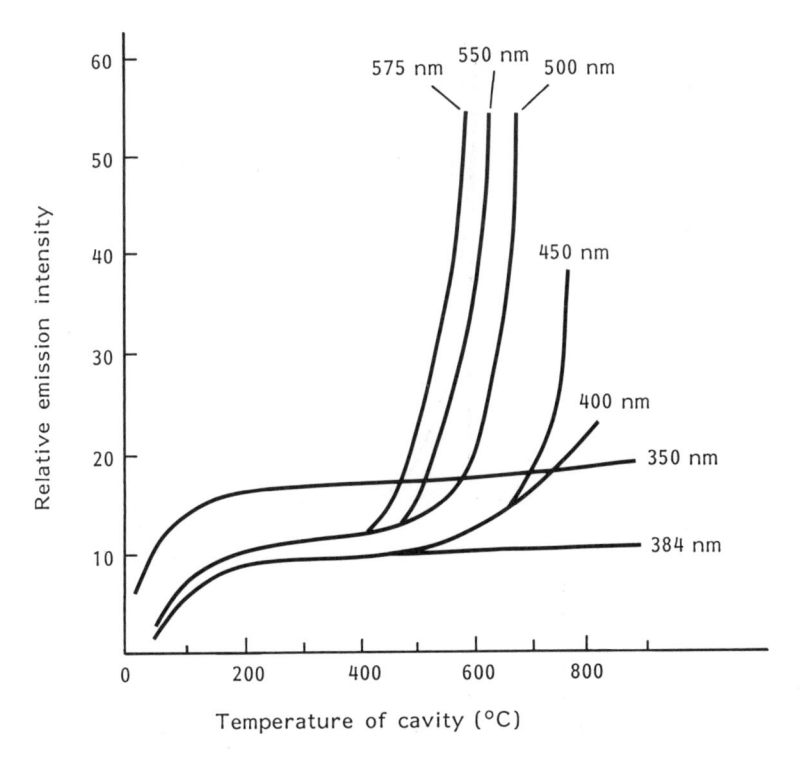

Figure 2.11. Incandescence emission profiles from a stainless-steel MECA cavity at different wavelengths when introduced into a hydrogen–nitrogen–air flame (45).

may be overcome by either using a cavity with low incandescence or by converting the analyte into a more volatile form. Thus, in the measurement of lecithin via its HPO emission at 528 nm where a high t_m value leads to greatly increased incandescence, a high density carbon cavity with a pyrolytic surface and a lower degree of incandescence than stainless-steel or ordinary carbon reduces the effect.

Incandescence is a major problem in the MECA determination of boron. The best way of overcoming this difficulty is to convert the boron compound into a more volatile product which can chemiluminesce before the cavity incandesces (see Chapter 5).

Cavity incandescence is not a problem with gas generation MECA detection using water-cooled cavities.

2.7.2 The Importance of the Salet Phenomenon in MECA

The ability of the MECA cavity to act as a cool body was first shown by

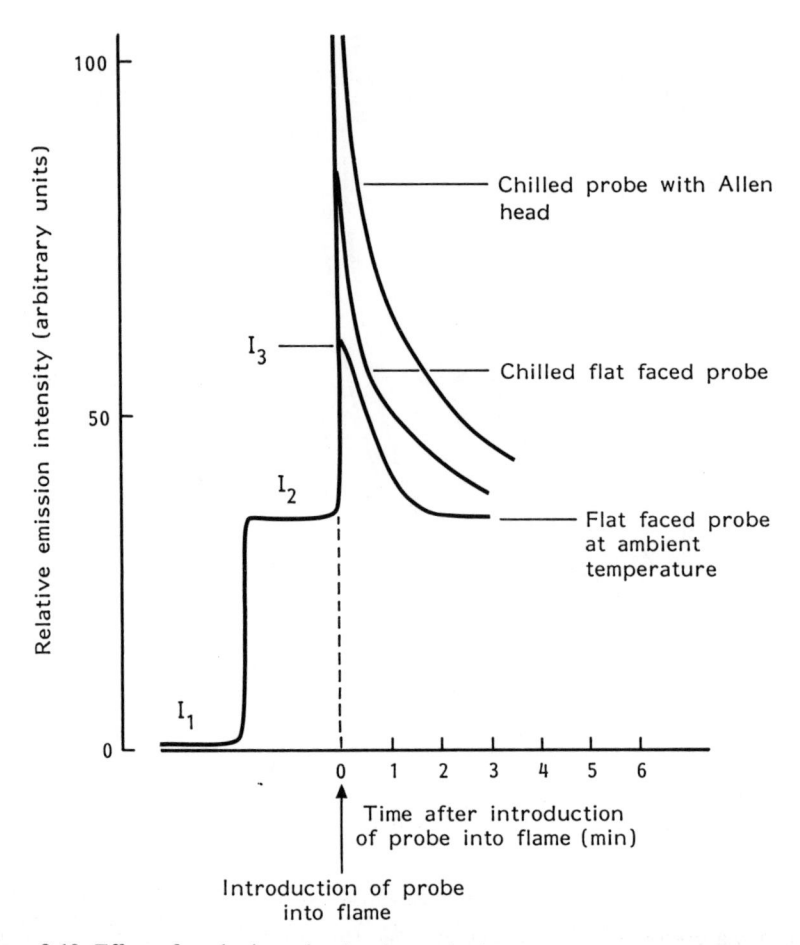

Figure 2.12. Effect of probe introduction into a hydrogen–nitrogen flame into which a sulphur compound is aspirated to generate S_2 emission (I_1: flame background emission, I_2: S_2 emission profile from aspirated aqueous thiourea; I_3: increase of S_2 emission intensity due to introduction of flat-faced probe at ambient temperature into the flame) (51).

Bogdanski (51) with an experiment similar to that performed by Salet. An aqueous ammonium sulphate solution was aspirated into a hydrogen diffusion flame and the blue S_2 emission was measured at 384 nm. An Allen screw, filed down to give a smooth flat surface, was introduced into the flame like an ordinary cavity. Immediately after introduction, an increase in the S_2 emission intensity was recorded which gradually decreased as the screw heated up (Figure 2.12). The increase in emission intensity was more marked if the probe was chilled prior to being introduced into the flame. If a cavity was used instead of a flat surface, a substantial increase in emission intensity was again

MOLECULAR EMISSION CAVITY ANALYSIS

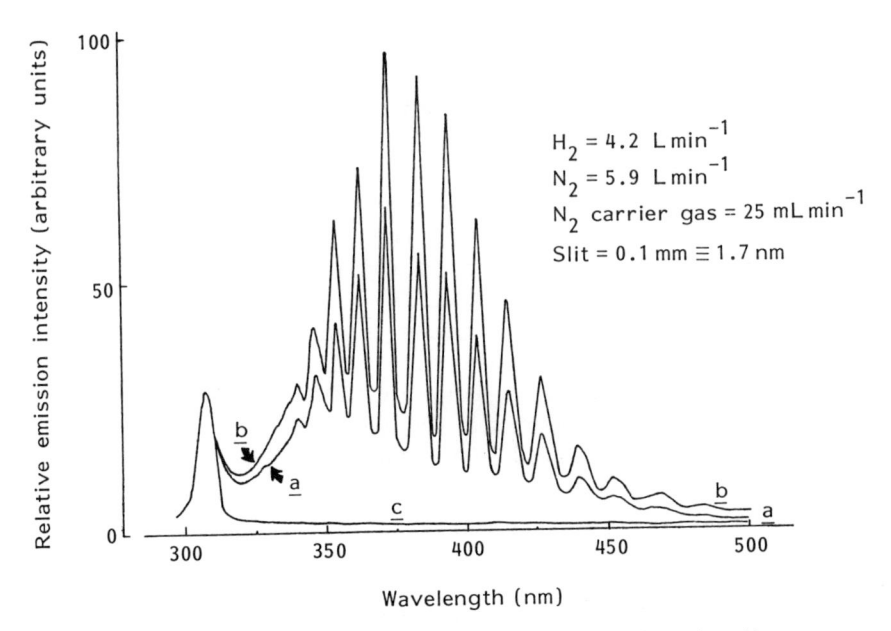

$H_2 = 4.2$ L min^{-1}
$N_2 = 5.9$ L min^{-1}
N_2 carrier gas = 25 mL min^{-1}
Slit = 0.1 mm ≡ 1.7 nm

Figure 2.13. S_2 emission spectra from carbon disulphide vapours introduced into a water-cooled MECA cavity into which a) 5 and b) 100 mL min^{-1} of water is supplied (c: flame background emission) (49).

observed, and the emitting species were confined to the cavity itself. Similar observations at 528 nm were made when a phosphorus compound was aspirated into the flame (51).

The enhancement of the S_2 emission intensity was further investigated by using a water-cooled cavity. The temperature of the cavity was altered by changing the flow rate of the cooling water and was monitored by a chrome–alumel thermocouple protruding into the cavity (53). The S_2 emission increased with an increasing flow rate of cooling water but the wavelength and relative intensities of the various bands did not change (Figure 2.13). It was concluded that the enhancement in emission intensity was not through any change in the nature of the emitting species nor to any re-distribution of energy among the various transitions.

Further studies by Shakir (54) revealed the presence of a sulphur deposit on the inner surface of the cavity thought to be the S_8 allotrope of sulphur (55). Sensitivity was also increased by treating the cavity with high amounts of sulphur-containing analyte prior to measurements being made. This treatment is often referred to as *conditioning the cavity* and is a necessary step in preparing any conventional MECA cavity for reproducible measurements. Similar effects have been observed in studies on Se_2 and HPO emissions.

The emission intensity from excited species which are formed within an oxy-cavity is reduced with increasing temperature. Typical examples of these species are CH, C_2, CN and BO_2 which are called *hot cavity emitters*. These emitters require oxygen either for their decomposition (e.g. CH, C_2, CN), or for their formation (e.g. BO_2), or because oxygen is necessary for a chemiluminescence reaction to take place within the cavity (e.g. $NO + O \longrightarrow NO_2 + h\nu$).

The exact mechanism of the Salet phenomenon and the effect of temperature on the hot cavity emitters is not fully understood. It appears that if the radiation is generated purely by chemiluminescent reactions, then a decrease in temperature increases the overall efficiency of CL. On the other hand, if thermal excitation is also involved, an increase in temperature produces a higher population of emitting species. These two mechanisms would, therefore, appear to work against one another. However, the temperature of the cavity can be controlled by the degree of water cooling selected. This unique feature of MECA allows the cavity temperature to be adjusted for optimal emission intensity of a given emitting species and introduces a considerable measure of selectivity into the technique.

2.7.3 Sample Introduction into the Cavity

Conventional MECA

The most important feature of conventional MECA is the t_m value, which allows speciation and determination of components from mixtures of compounds which contain the same element. For compounds which vaporize by boiling (e.g. many organic sulphur compounds and sulphuric acid), sublimation (e.g. selenium dioxide) and by thermal decomposition (e.g. thiourea, sulphates of iron(II), manganese(II) and sodium), the t_m value is directly related to their boiling, sublimation and thermal decomposition temperatures. The order in which emissions from these compounds appear does not depend on the cavity material, position in the flame, the degree of external cooling, or flame composition. The t_m value and the emission intensity, however, change with the heating rate of the cavity.

The effect of experimental parameters on the t_m value has been studied by focusing an optical pyrometer on the inner surface of the cavity (56). The effect of flame composition on the temperature profile of a stainless-steel cavity is shown in Figure 2.14. From this it is clear that if a compound vaporizes at 250°C, it has a t_m value of about 29 s using gas flow rates of 1.15 L min^{-1} (H_2), 5.0 L min^{-1} (N_2). Addition of air increases the temperature of the flame and reduces the t_m value (see Figure 2.9 for thiourea). The position of the cavity within the flame determines its rate of heating, so not unexpectedly this parameter also has an effect on the t_m value (Figure 2.15).

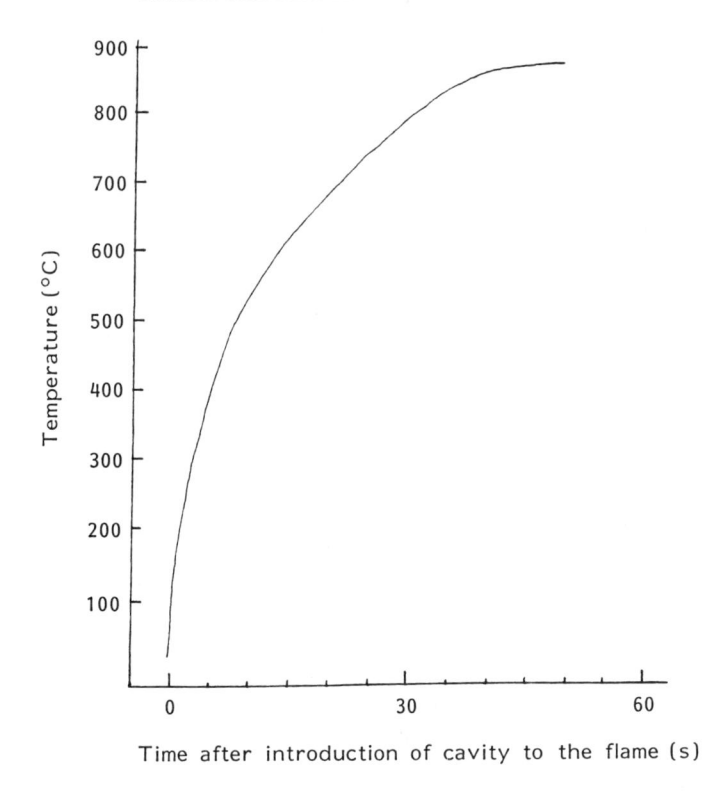

Figure 2.14. Typical temperature profile of stainless-steel MECA cavity after introduction into the flame (51).

Ceramic insulation between the cavity and holder in a stainless-steel cavity reduces the rate of heat loss and the cavity heats up faster. Therefore, a compound vaporizing at 700°C does not vaporize in a carbon cavity and no MECA response is observed because heat is conducted away too quickly. In a stainless-steel cavity, it has a t_m value in the region of 27 s, showing that the cavity retains sufficient heat for the compound to vaporize. When a ceramic insulator is incorporated into the cavity holder, the t_m value is reduced to about 17 s.

Compounds such as phenothiazines which vaporize by catalytic breakdown, have shorter t_m values in stainless-steel than in carbon cavities because of the catalytic effect of metals in the steel. The t_m value of a given compound is clearly affected by several experimental parameters. If these parameters are kept constant, then the t_m value can be used for speciation and quantitative work. However, care must be exercised because compounds which vaporize within ±50°C of each other do not show resolved responses.

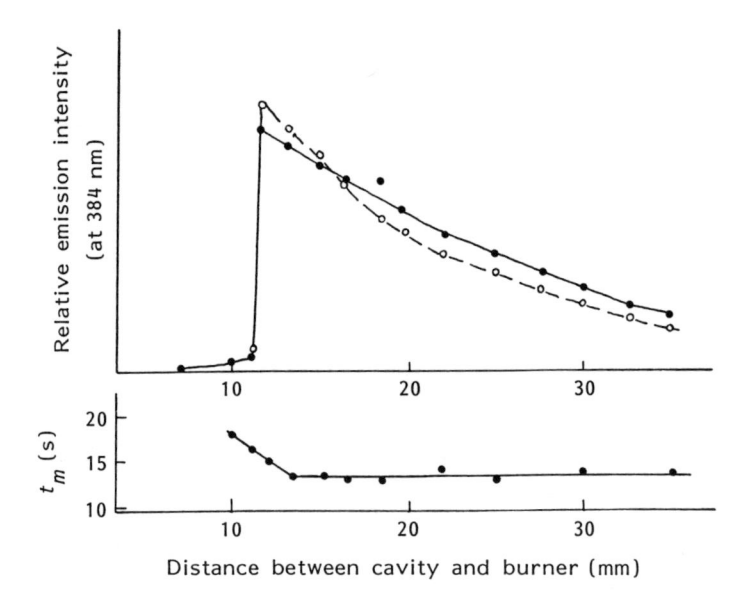

Figure 2.15. Effect of height of MECA cavity above burner top on the S_2 emission intensity and t_m from iron(II) sulphate (cavity orifice vertically above central axis of burner) (○) peak height, (●) integrated emission (45).

A good example of how the t_m value can be used to determine the components of a mixture is shown in Figure 2.16 where the S_2 emission responses from a mixture containing sulphide, sulphite and sulphate (or peroxodisulphate) is depicted. In the absence of air, sulphide and sulphite can be determined directly from the appropriate S_2 emission intensities which have well-resolved t_m values. The response from sulphate is obtained only after air is introduced into the hydrogen diffusion flame to vaporize the refractory metal sulphate (57).

An alternative to t_m is T_m, which is defined as *the temperature of the cavity at which the t_m value is reached* and does not depend on experimental conditions. Measurement of T_m is effected by drilling a hole along the centre of the threaded portion of the cavity to within about 1 mm of the cavity's wall (51). Thus, emission profiles can be measured as a function of temperature on an X–Y recorder, as shown in Figure 2.17 for a variety of sulphur compounds in a stainless-steel cavity. Although the T_m value has not been used for routine MECA work, its use minimizes difficulties associated with using t_m values.

Solvent and matrix effects. When conventional MECA is performed with solutions, the solvent may greatly affect the response if it evaporates at the same time as the emitting analyte. Organic solvents, for instance, quench S_2 emissions because they form thermodynamically stable gaseous carbon–sulphur species (58).

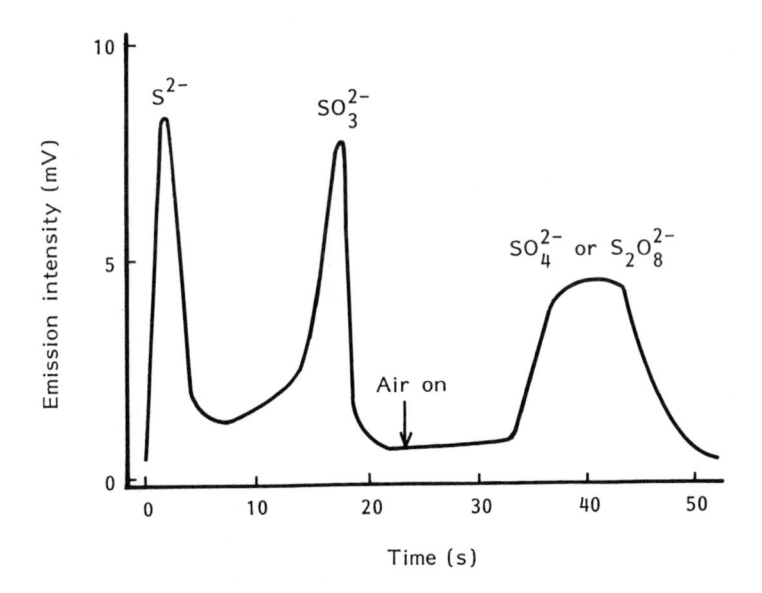

Figure 2.16. S_2 responses from 5 μL of a mixture containing 10 μg S mL^{-1} as Na$_2$S + 16 μg S mL^{-1} as Na$_2$SO$_3$ + 40 μg S mL^{-1} as Na$_2$SO$_4$ or Na$_2$S$_2$O$_8$ in 0.01 mol L^{-1} (NH$_4$)$_2$HPO$_4$ injected into a stainless-steel MECA cavity (gases: 1.7 L H$_2$ min^{-1}, 4.0 L N$_2$ min^{-1}, 1.5 L air min^{-1} was introduced into the flame as indicated) (57).

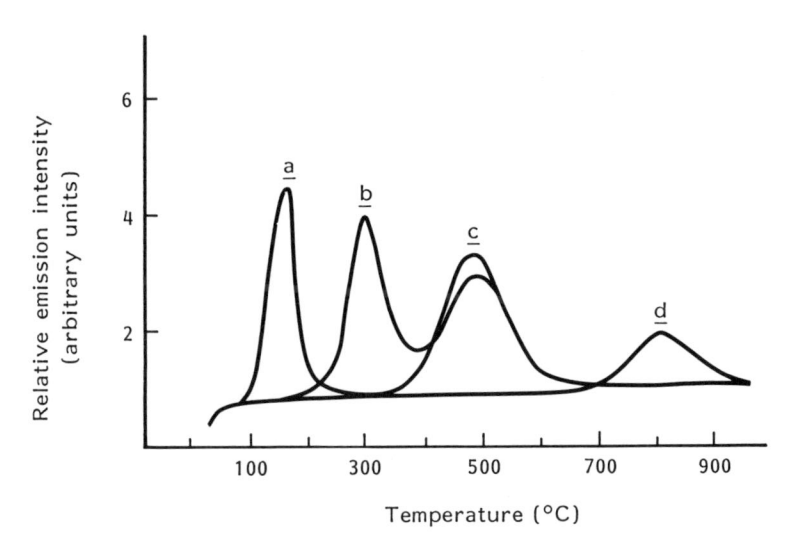

Figure 2.17. S_2 emission profiles at 384 nm as a function of T_m from (a) thiourea, (b) ammonium sulphate (also showing iron(II) sulphate emission due to iron), (c) iron(II) sulphate, and (d) sodium sulphate (51).

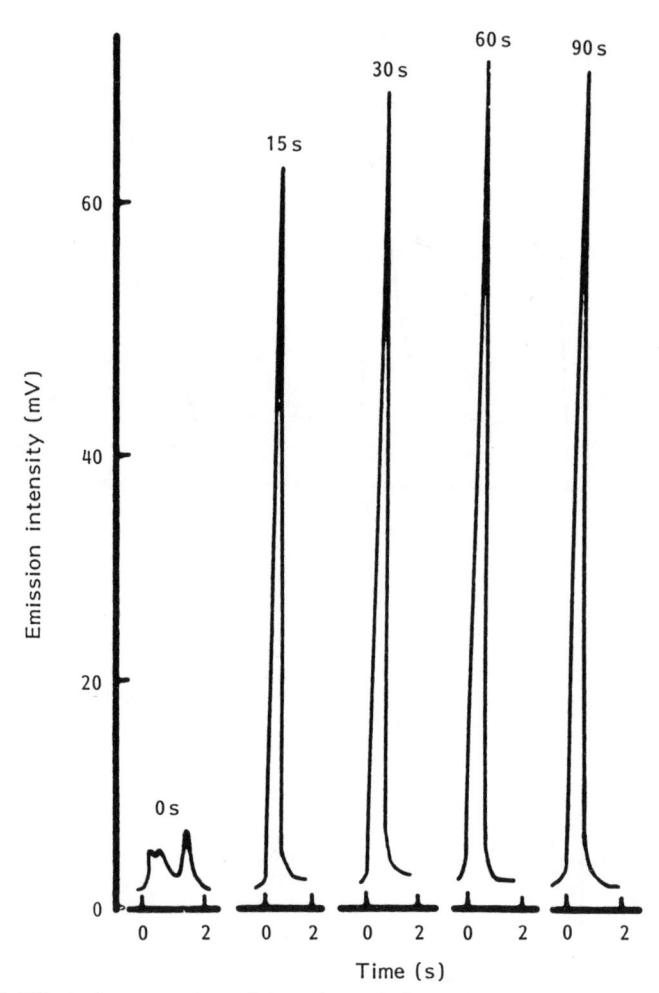

Figure 2.18. Effect of evaporation of the solvent (n-hexane) on the S_2 emission intensity from 10 ng of sulphur as promethazine in 5 μL aliquots, using a stainless-steel MECA cavity and a flame of 6.0, 6.0 and 4.5 L min^{-1} for hydrogen, nitrogen and air respectively (59).

Promethazine dissolved in n-hexane provides a typical example of how a solvent affects the molecular emission profile (Figure 2.18) (59). Hexane normally vaporizes at the same time as the promethazine in the cavity. When desulphurization of promethazine takes place, hydrogen sulphide is evolved so that the cavity space is rich in hydrocarbon vapours and the S_2 emission is severely quenched. However, if the hexane is evaporated before the cavity is heated, then an intense and sharp S_2 response is recorded from the promethazine.

Water, as a solvent, also affects the response from promethazine. If water is present in the cavity, it vaporizes about 1 s after introduction into the flame and since promethazine vaporizes after 0.2 s, the resulting response is broad and delayed ($t_m = 2.4$ s). The response is sharp if water is evaporated before introduction of the cavity into the flame.[1] It appears that water reacts with hydrogen atoms in the cavity, reducing their concentration, and thus affecting the emission response:

$$H + H_2O \longrightarrow OH\cdot + H_2 \qquad (28)$$

When samples containing relatively non-volatile compounds are injected into the cavity, a build-up of solid deposit may occur. Cavity reflectivity and size, and the rate of heat removal from the cavity to the holder change and the cavity must be discarded. Residue formation may be overcome to some extent by using a silica-lined cavity or a carbon cavity. Matrix effects lead to considerable problems in conventional MECA (60) and are a prime cause for its general lack of applicability.

Multi-peaked responses sometimes appear if a sulphur compound like methionine (250 ng S) or 1,3-diethylthiourea (100 ng S) are injected into a silica cavity. Nakajima *et al.* (61), for instance, have found that the cavity must be conditioned before being used, by injecting into it a large amount of alkali metal chloride and then heating the cavity inside the flame. After the conditioning procedure, the cavity surface becomes rough and promotes sample dispersion. Thus, peak splitting is avoided and sharp peaks are recorded.

The major disadvantage of conventional MECA is that the Salet enhancement of S_2, Se_2 and HPO emissions cannot be fully utilized. As the cavity heats up, Salet participation in the response is decreased and, therefore, the emission intensity depends not only on time but on temperature as well. Thus, emission responses cannot be integrated with respect to time and peak areas cannot be used for quantitative work.

Gas Generation MECA Detection

Problems associated with solvent and matrix effects within the cavity can be avoided by *gas generation MECA detection*. The sample is converted into a gas or a vapour which is introduced into the cavity. The excited molecules are formed almost instantaneously and the emission appears very quickly thereafter. The t_m concept is not valid and components of a mixture can be determined simultaneously only if they are resolved prior to introduction into the cavity, i.e. by a series of chemical reactions or through separation on a gas chromatographic column.

---

[1]The most common and simple way of evaporating water prior to cavity introduction into the flame is by keeping the cavity upwards and blowing hot air towards the injected solution for a preselected period of time which must be reproducibly controlled.

The major advantages of gas generation MECA detection are:

1. The cavity is not used for sample vaporization so water-cooled cavities can be employed for effective use of the Salet phenomenon.
2. A great variety of chemical reactions can be used.
3. The technique can be very easily automated for flow injection or continuous flow manifolds.

Gas generation MECA detection can be achieved by one of the following methods of gas evolution:

1. Chemical methods. Chemical methods use a selective chemical reaction to convert the analyte into a vapour which is then carried into the cavity by an inert gas to form the emitting species. The selectivity, sensitivity and limit of detection depend primarily on the chemical reaction and the design of the gas generation system. The chemical reaction must fulfil the following major requirements:

1. The reaction must be simple, requiring small amounts of one or, at most, two reagents.
2. The reaction must be fast so that the vapours can be carried directly into the cavity for high response. Slow reactions are responsible for low and broad peaks which can still be used after integration but the measurement time is lengthened. In some cases, it is convenient to collect the vapours (e.g. by liquid nitrogen trapping techniques) and introduce them instantaneously into the cavity.
3. The reaction must be stoichiometric without any volatile by-products which might emit within the cavity.
4. The final reaction solution must not be so viscous that it causes delay or retainment of vapours. If a viscous reagent such as orthophosphoric acid is used, elevated temperatures are required for the reaction. Chemical methods can be classified according to the reaction which generates the vapour:

(a) *Hydride generation methods*. The best reagent for hydride generation is sodium borotetrahydride or tetrahydroborate ($NaBH_4$) which is used for the generation of volatile hydrides such as AsH_3, SbH_3, SnH_4, H_2Se and H_2Te. The reagent has been used as solid (62, 63), or as a 2–4% m/v solution in 0.01 mol L^{-1} sodium hydroxide (64). Selective hydride generation of arsenic and selenium can be achieved by initially adjusting the acidity of the reaction solution to 0.1 mol L^{-1} in HCl to generate arsine, and then liberating hydrogen selenide by increasing the acidity to 2 mol L^{-1}.

(b) *Oxidation-reduction methods*. These methods include: (i) *reduction of inorganic and organic sulphates and sulphonates* to hydrogen sulphide after reaction with tin and condensed phosphoric acid at elevated temperatures (65); (ii) *generation of nitrogen monoxide from nitrite ions* with iodide in acidic solutions

$$2HNO_2 + 2I^- + 2H^+ \longrightarrow I_2 + 2NO + 2H_2O \qquad (29)$$

(iii) *reduction of nitrate and nitrite to nitrogen monoxide* with granular zinc; (iv) *reduction of nitrate to ammonia* with Devarda's alloy in alkaline medium (66).

(c) *Acidic generation methods*. Concentrated or aqueous solutions of a strong acid are used to generate the analyte vapour. Examples are the generation of hydrogen sulphide from sulphide solutions (67) or sulphur dioxide from sulphite solutions (68, 69) with orthophosphoric acid, generation of carbon dioxide from carbonate with hydrochloric acid (70), and generation of nitrogen monoxide from nitrite with concentrated orthophosphoric acid.

(d) *Alkaline generation methods*. The only reported example of these methods is the evolution of ammonia from ammonium ions by treatment with sodium hydroxide pellets (66).

(e) *Miscellaneous generation methods*. These methods include the formation of methyl borate from boron compounds by treatment with a mixture of methanol–sulphuric acid, and formation of silicon tetrafluoride from inorganic silicon compounds by the action of excess fluoride and concentrated sulphuric acid (71). Also, thiosemicarbazide and dithioxamide have been determined by electrolytic generation of hydrogen sulphide (72). This method seems feasible for the determination of arsenic, antimony and silicon compounds and requires further investigation.

2. Gas chromatographic methods. The selectivity, versatility and resemblance of the cavity to the flame photometric detector (FPD) has encouraged an evaluation of MECA as a detector for gas chromatography. The advantage of MECA over the FPD is that the cavity can be used for a greater variety of elements and can be proposed as a universal detector for most non-metals and metalloids as well as for some metals.

3. Physical methods. Refractory compounds or compounds which cannot be converted chemically into a vapour for MECA detection, can be converted by an electrically-heated tantalum filament (Figure 2.19) (73). The filament is positioned inside a glass container which is continuously purged by nitrogen. The sample is injected onto the filament which is then heated so that the volatile analyte may be swept into the cavity. A modified Massmann furnace has also been used (Figure 2.20) (74). Gas inlet and outlet ducts are attached to the ends of the furnace, through which hydrogen enters and leaves the interior of the graphite tube. Nitrogen is retained as the shielding gas for the other parts of the furnace. Both vapour generation techniques have been evaluated for a wide range of compounds (75).

The limited research work on electrically generated vapours indicate that MECA can easily be used as a detector for gases evolved from the controlled environment of a thermobalance or any other thermal analytical device.

2.7.4 Molecular Emissions and Spectra

MECA emissions give either band or continuous spectra. Band spectra are

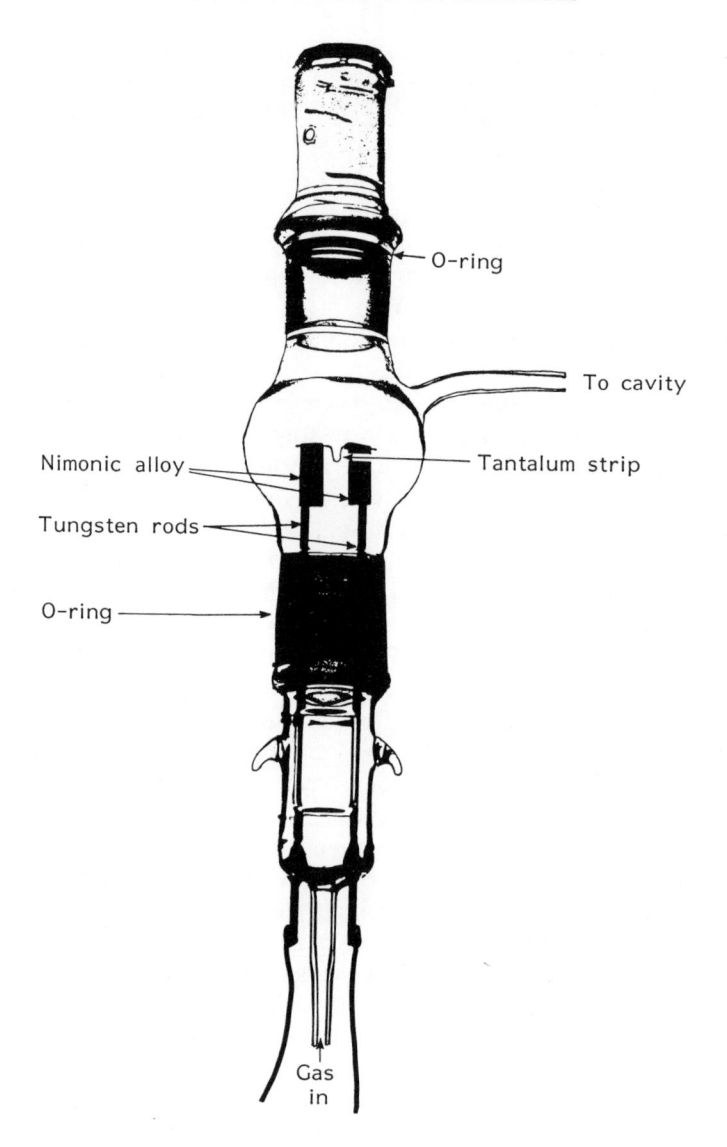

Figure 2.19. Electrically heated tantalum filament (73).

characterized by transitions between quantized energy states while continuous spectra appear when the energy states are so close together as to appear unquantized (76). Continuous spectra are generated by various chemical processes such as the association process:

$$AsO + O \longrightarrow AsO_2 + h\nu \tag{30}$$

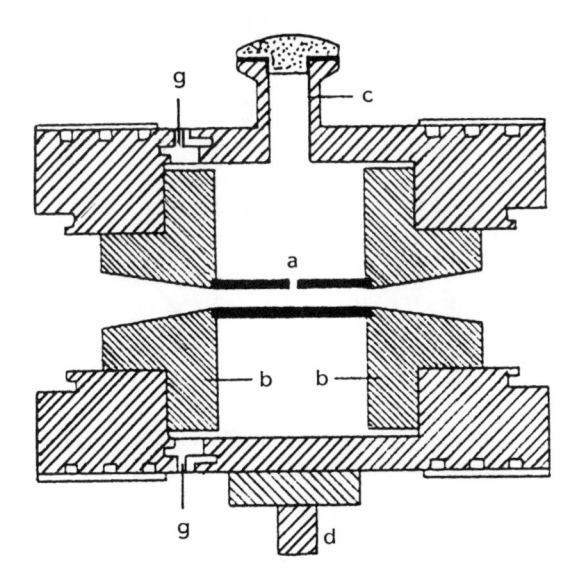

Figure 2.20. Modified Massmann furnace (74); (a) Graphite tube, (b) steel flanges, (c) sample introduction port, (d) mount, and (g) plastic insulator.

The S_2 emission spectrum (Figure 2.13) is a typical band spectrum with maximum intensities at 384 and 394 nm. The former is mostly used for quantitative MECA studies. This particular emission is extremely sensitive, being visible to the human eye when only a few $\mu g\ mL^{-1}$ sulphur are vaporized. It is the most sensitive emission for MECA studies of sulphur compounds and its magnitude is further increased by the Salet phenomenon.

The mechanism of S_2 formation involves a reaction between two sulphur atoms to generate one photon. Hence, theoretically, the emission intensity (I) is directly proportional to the square of sulphur concentration, C:

$$I \cong [S_2^*] = KC^n \tag{31}$$

where K is the reaction constant. The maximum theoretical value of n is 2 but experimental values range between 1 and 2. The exponential relation of emission intensity to analyte sulphur concentration is responsible for the parabolic shape of the calibration graph (Figure 2.21). The curve is linearized by plotting log I vs log C from which the value of n and K can also be found

$$\log I = \log K + n \log C \tag{32}$$

Negative deviations are observed at high concentrations of sulphur. These deviations can be attributed to self-absorption or to the formation of other polyatomic sulphur molecules, such as S_8, which emit at other wavelengths or do not emit at all. The most probable explanation for the negative deviation

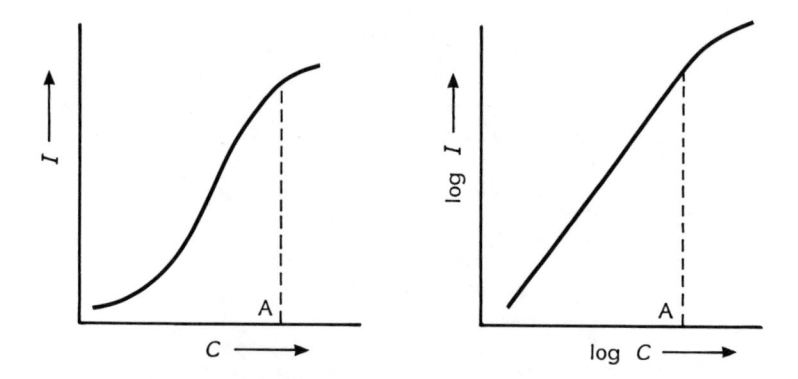

Figure 2.21. Typical calibration graph of S_2 emission intensity (I) *versus* concentration of sulphur (C) and linearization by plotting the log I *versus* log C curve showing negative deviations after A.

is self-absorption for which the following relationship has been proposed (77)

$$I = aC^2 10^{-bC^2} \tag{33}$$

where a and b are constants and the factor 10^{-bC^2} corrects for self-absorption.

The relative standard deviations for I and C are related by (78):

$$S_C/C = (1/n)(S_I/I) \tag{34}$$

where S_C and S_I are the standard deviations for C and I, respectively. The sensitivity of the method is (79):

$$dI/dC = nKC \tag{35}$$

and therefore, there is no physical meaning for the limit of detection (i.e. as $C \to 0, dI/dC \to 0$) and sensitivity rapidly increases with concentration. Since the formation of excited S_2 from sulphide is simpler than that from sulphite, the value of n for the former is higher than for the latter (80). Measurement of sulphide is more precise and sensitive than of sulphite (81). Hence, the most sensitive method for determining a sulphur compound is by gas generation of hydrogen sulphide.

Other spectra will be discussed in individual chapters.

2.8 CONCLUSIONS

Since the discovery of MECA, the advantages of chemiluminescence have been fully investigated and established. The sensitivity of gas phase CL especially for sulphur-containing compounds which give the S_2 emission, is now well proven. However, flame CL does have disadvantages which relate primarily to the temperature of the flame and the inability to take full advantage of the Salet phenomenon. Clearly, it would be a great step forward if a MECA-based method, dependent for its CL energy on a non-flame hydrogen atom recombination process, could be developed.

REFERENCES

1. B. Radziszewski, *Ber.*, **10**, 70 (1877).
2. H. O. Albrecht, *Z. Phys. Chem.*, **136**, 321 (1928).
3. K. Robards and P. J. Worsfold, *Anal. Chim. Acta*, **266**, 147 (1992).
4. W. R. G. Baeyens, D. De Keukeleire, and K. Korkidis, Eds., *Luminescence Techniques in Chemical and Biochemical Analysis*, Marcel Dekker, New York, NY, 1991.
5. J. H. Glover, *Analyst.*, 100, 449 (1975).
6. M. Katz, Ed., *Methods of Air Sampling and Analysis*, 2nd ed., American Public Health Association, Washington, DC, 1977.
7. M. E. Mulder, *Bull Soc. Chim. France*, **1**, 453 (1864).
8. W. F. Barrett, *Phil. Mag.*, **30**, 321 (1865).
9. G. Salet, *Bull. Soc. Chim. France*, **11**, 302 (1869).
10. G. Salet, *Compt. Rend.*, **68**, 404 (1869).
11. A. G. Gaydon, *The Spectroscopy of Flames*, 2nd ed., Chapman and Hall, London, 1974.
12. A. G. Gaydon and H. G. Wolfhard, *Flames, Their Structure, Radiation and Temperature*, 3rd ed., Chapman and Hall, London, 1970.
13. T. M. Sugden and A. Demerdache, *Nature*, **195**, 596 (1962).
14. S. A. Ghonaim, *Proc. Soc. Anal. Chem.*, **11**, 167 (1974).
15. G. Salet, *Bull Soc. Chim. France*, **14**, 182 (1870).
16. G. Salet, *Bull. Soc. Chim. France*, **13**, 289 (1870).
17. G. Salet, *Compt. Rend.*, **73**, 1056 (1871).
18. G. Salet, *Bull. Soc. Chim. France*, **16**, 195 (1871).
19. W. L. Crider, *Anal. Chem.*, **37**, 1770 (1965).
20. R. M. Dagnall, K. C. Thompson, and T. S. West, *Analyst.*, **92**, 506 (1967).
21. R. M. Dagnall, K. C. Thompson, and T. S. West, *Analyst.*, **93**, 72 (1968).
22. A. Syty and J. A. Dean, *Appl. Opt.*, **7**, 1331 (1968).
23. G. M. Trischan and C. W. Frank, *Abstract No. 65, Pittsburgh Conference on Analytical Chemistry and Applied Spectroscopy, Cleveland, OH, February 28–March 4, 1977.*
24. K. G. Brodie, *Resonance Lines*, vol. 1, no. 3, Varian Techtron, Walnut Creek, CA, 1969.
25. F. P. Scaringelli and K. A. Rehme, *Anal. Chem.*, **41**, 707 (1969).
26. K. M. Aldous, R. M. Dagnall, and T. S. West, *Analyst*, **95**, 1130 (1970).
27. C. Veillon and J. Y. Park, *Anal. Chim. Acta,* **60**, 293 (1972).
28. R. S. Braman, J. M. Ammons, and J. L. Bricker, *Anal. Chem.*, **50**, 992 (1978).
29. A. Hadjitofi and J. G. Wilson, *Atmos. Environ.*, **13**, 755 (1979).
30. J. F. Alder and K. Kargosha, *Anal. Chim. Acta*, **107**, 231 (1979); **111**, 145 (1979).
31. W. N. Elliott and R. A. Mostyn, *Analyst.*, **96**, 452 (1971).
32. W. N. Elliott, C. Heathcote, and R. A. Mostyn, *Talanta.*, **19**, 359 (1972).
33. Y. Y. Cheung, G. F. Kirkbright, and R. D. Snook, *Anal. Chim. Acta.*, **140**, 213 (1982).

34. E. E. Pickett, J. C.-M. Pau, and S. R. Koirtyohann, *J. Assoc. Off. Anal. Chem.*, **54**, 796 (1971).
35. J. M. S. Butcher and G. F. Kirkbright, *Analyst.*, **103**, 1104 (1978).
36. P. T. Gilbert, *Anal. Chem.*, **38**, 1920 (1966).
37. R. M. Dagnall, K. C. Thompson, and T. S. West, *Analyst.*, **94**, 643 (1969).
38. R. M. Dagnall, K. C. Thompson, and T. S. West, *Analyst.*, **93**, 518 (1968).
39. R. M. Dagnall, D. J. Smith, K. C. Thompson, and T. S. West, *Analyst.*, **94**, 871 (1969).
40. R. M. Dagnall, B. Fleet, T. H. Risby, and D. R. Deans, *Talanta.*, **18**, 155 (1971).
41. S. S. Brody and J. E. Chaney, *J. Gas Chromatogr.* **4**, 42 (1966).
42. M. C. Bowman and M. Beroza, *Anal. Chem.*, **40**, 1448 (1968).
43. T. L. Chester, *Anal. Chem.*, **52**, 638, 1621 (1980).
44. R. Belcher, S. L. Bogdanski, S. A. Ghonaim, and A. Townshend, *Anal. Lett.*, **7**, 133 (1974).
45. R. Belcher, S. L. Bogdanski, and A. Townshend, *Anal. Chim. Acta*, **67**, 1 (1973).
46. R. Belcher, T. Kouimtzis, and A. Townshend, *Anal. Chim. Acta*, **68**, 297 (1974).
47. M. Q. Al-Abachi, *Proc. Anal. Div. Chem. Soc.*, **14**, 251 (1977).
48. T. Sugiyama, Y. Suzuki, and T. Takeuchi, *J. Chromatogr.*, **77**, 309 (1973); **85**, 45 (1973).
49. A. C. Calokerinos, PhD Thesis, University of Birmingham, UK, 1978.
50. O. Osibanjo, PhD Thesis, University of Birmingham, UK, 1976.
51. S. L. Bogdanski, PhD Thesis, University of Birmingham, UK, 1973.
52. R. Belcher, S. L. Bogdanski, A. C. Calokerinos, and A. Townshend, *Analyst.*, **102**, 220 (1977).
53. S. L. Bogdanski, A. C. Calokerinos, and A. Townshend, *Can. J. Spectrosc.*, **27**, 10 (1982).
54. I. M. A. Shakir, PhD Thesis, University of Birmingham, UK, 1980.
55. A. C. Calokerinos and T. P. Hadjiioannou, *Anal. Chim. Acta*, **148**, 277 (1983).
56. I. H. B. Rix, Ph. D. Thesis, University of Birmingham, UK, 1976.
57. M. Q. Al-Abachi, R. Belcher, S. L. Bogdanski, and A. Townshend, *Anal. Chim. Acta*, **86**, 139 (1976)
58. S.-A. Fredriksson and A. Cedergen, *Anal. Chim. Acta*, **100**, 429 (1978).
59. M. Q. Al-Abachi, PhD Thesis, University of Birmingham, UK, 1977.
60. R. Ajlek and J. Stupar, *Vestn. Slov. Kem. Drus.*, **33**, 87 (1986).
61. K. Nakajima, K. Ohta, and T. Takada, *Anal. Chim. Acta*, **270**, 247 (1992).
62. R. Belcher, S. L. Bogdanski, S. A. Ghonaim, and A. Townshend, *Anal. Chim. Acta*, **72**, 183 (1974).
63. R. Belcher, S. L. Bogdanski, E. Henden, and A. Townshend, *Anal. Chim. Acta*, **92**, 33 (1977).
64. R. Belcher, S. L. Bogdanski, E. Henden, and A. Townshend, *Anal. Chim. Acta*, **113**, 13 (1980).
65. S. L. Bogdanski, I. M. A. Shakir, W. I. Stephen, and A. Townshend, *Analyst.*, **104**, 886 (1979).
66. R. Belcher, S. L. Bogdanski, A. C. Calokerinos, and A. Townshend, *Analyst.*, **106**, 625 (1981).
67. N. Grekas and A. C. Calokerinos, *Anal. Chim. Acta*, **173**, 311 (1985).
68. S. L. Bogdanski, A. Townshend, and B. Yenigul, *Anal. Chim. Acta*, **115**, 361 (1980).
69. A. C. Calokerinos and A. Townshend, *Fresenius Z. Anal. Chem.*, **311**, 214 (1982).
70. S. L. Bogdanski, E. Henden, and A. Townshend, *Anal. Chim. Acta*, **116**, 93 (1980).
71. M. Burguera, A. Townshend, and S. L. Bogdanski, *Anal. Chim. Acta*, **117**, 247 (1980).

REFERENCES

72. N. Grekas and A. C. Calokerinos, *Anal. Chim. Acta*, **202,** 241 (1987).
73. N. Pourezza, PhD Thesis, University of Birmingham, UK, 1981.
74. P. S. Turner, PhD Thesis, University of Hull, UK, 1986.
75. N. Pourezza, A. Townshend, and P. S. Turner, *Anal. Proc.* **25,** 244 (1988).
76. A. G. Gaydon, *Dissociation Energies and Spectra of Diatomic Molecules*, 3rd ed., Chapman and Hall, London, 1968.
77. D. G. Greer and T. J. Bydalek, *Environ. Sci. Technol.*, **7,** 153 (1973).
78. J. C. Miller and J. N. Miller, *Statistics for Analytical Chemistry*, 2nd ed., Horwood, Chichester, 1988.
79. Analytical Methods Committee, *Analyst.*, **45,** 532 (1987).
80. N. Grekas, PhD Thesis, University of Athens, Greece, 1988.
81. N. Grekas and A. C. Calokerinos, *Anal. Chim. Acta*, **225,** 359 (1989).
82. R. Belcher, S. L. Bogdanski, D. J. Knowles, and A. Townshend, *Anal. Chim. Acta.*, **77,** 53 (1975).

CHAPTER

3

INSTRUMENTATION AND AUTOMATION

NIKOLAOS GREKAS

Research and Development Laboratory
Farmalex S.A.
Athens, Greece

3.1	**Introduction**		43
3.2	**Instrumentation**		44
	3.2.1	Emission Burner Unit	44
	3.2.2	Cavity Probe and Holder Unit	45
		3.2.2.1 Cavity Probe	45
		3.2.2.2 Cavity Holder	50
	3.2.3	Optical Unit – Readout System	53
	3.2.4	Gas Generation Systems	54
3.3	**Commercial Instruments**		55
	3.3.1	Conventional MECA	55
		3.3.1.1 MECA–22 Spectrometer	55
	3.3.2	Gas Generation Detection	56
		3.3.2.1 MECA–VAP	56
		3.3.2.1.1 Design and Characteristics of MECA–VAP	57
		3.3.2.1.2 Design and Characteristics of the MEP–101/ DIVAP–201	58
	3.3.3	Gas and High Performance Liquid Chromatographic Detection	59
3.4	**Automation**		61
	3.4.1	Conventional Automated Analysers	63
	3.4.2	Automated Gas Generation Analysers	65
3.5	**Conclusions**		69
References			69

3.1 INTRODUCTION

MECA has been used on its own (conventional MECA) or as a detector for vapour generating systems (MECA–VAP). Research has also been devoted to the development of MECA detectors for gas and liquid chromatography.

Both basic MECA techniques use the same instrumentation but in MECA–

Flame Chemiluminescence Analysis by Molecular Emission Cavity Detection
Edited by D. A. Stiles, A. C. Calokerinos and A. Townshend   © 1994 John Wiley & Sons Ltd

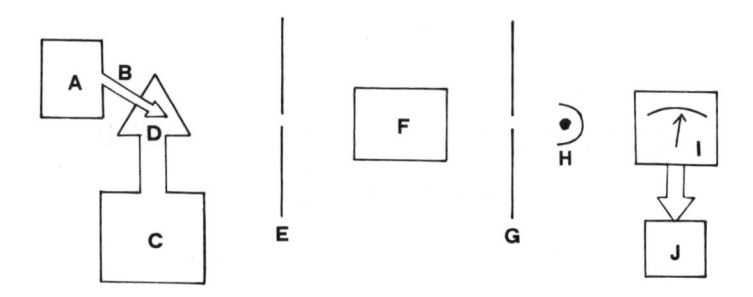

Figure 3.1. The essential parts of a MECA spectrometer (not to scale). A: cavity holder; B: probe and cavity; C: burner; D: flame; E: entrance slit; F: wavelength selector; G: exit slit; H: detector; I: readout device; J: recorder.

VAP an appropriate chemical reactor, is necessary to generate gaseous analyte by a chemical reaction. This reactor operates in conjunction with the cavity.

3.2 INSTRUMENTATION

The major components of a MECA spectrometer are the burner, the cavity probe and holder, the optical unit, and the readout system (Figure 3.1). The only component that is unique and, therefore, requires special design and careful construction is the cavity probe. A MECA instrument can be designed by simple modification of a commercial flame emission spectrometer (FES), or an atomic absorption spectrophotometer (AAS) operating in the emission mode. For this reason, common FES/AAS instruments such as the EEL 240 MK II (Evans Electroselenium Ltd) and the Pye Unicam SP 900 have been used for MECA measurements. Since the instrumentation is simple and readily available, most of the MECA instruments used by research teams during the period 1972–1990 were laboratory assembled. These spectrometers were each constructed in separate laboratories using component parts from commercial AAS or FES instruments.

Commercial MECA instruments were developed by two different manufacturers. These are the MECA-22 spectrometer (Anacon, Inc., Houston TX, USA) and the MEP-101 (Doma Instruments, Sherwood Scientific Ltd, Science Park, Milton Road, Cambridge, UK) MECA device. These instruments will be described later.

3.2.1 Emission Burner Unit

A typical commercially available emission burner (15 mm i.d., 150 mm height) is normally used. High purity hydrogen and nitrogen are supplied to the burner

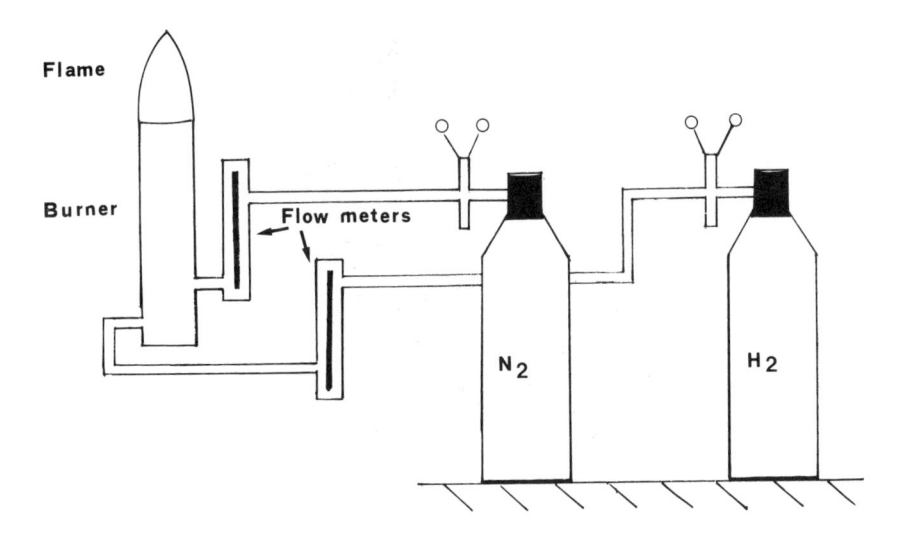

Figure 3.2. Typical emission burner used in MECA.

through the appropriate inlets at flow rates of about 4.5 and 6.5 L min^{-1}, respectively (Figure 3.2). If a third gas, such as oxygen or air is also required, it is introduced through an auxiliary inlet (1). All gases are supplied from gas cylinders and can be mixed prior to entering the burner. Gas pressures and flow rates are monitored by the appropriate pressure controllers and flow meters respectively. The burner is positioned perpendicularly to the axis of the optical system, a few centimetres from the entrance slit. Other modified burners have also been used in order to achieve maximum emission intensity and therefore maximum sensitivity (2), lower background emission from the flame, and less consumption of flame gases (3).

3.2.2 Cavity Probe and Holder Unit

3.2.2.1 Cavity Probe

Even though all other components of a MECA spectrometer are readily available, the cavity probe and holder units must be specially designed. However, they are easily constructed and can be arranged to fit into any common commercial flame photometer or any other laboratory-assembled MECA instrument.

The cavity is the most important component of the instrument and differentiates MECA from other related techniques (see Chapter 2). A small aperture (hole) is cut at the end of a metallic or carbon rod (probe) which is mounted on a suitable holder (Figure 3.3). The cavity defines a small area of the flame within

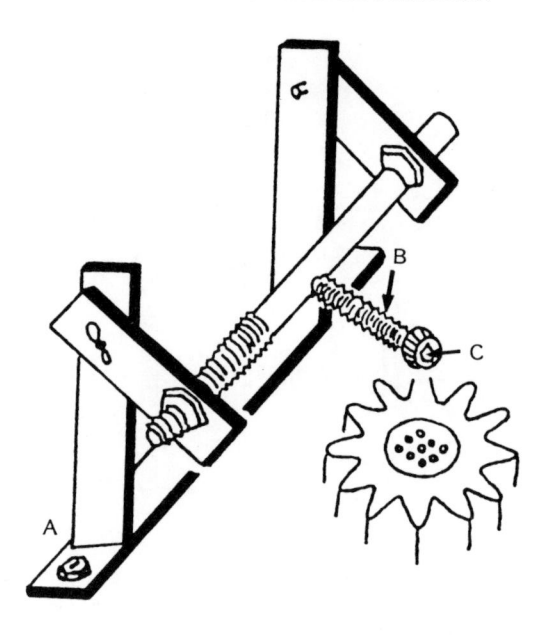

Figure 3.3. Cavity and cavity holding system used in early MECA experiments: (A) holder, (B) probe, (C) cavity.

which the excited molecules are formed and stabilized. It, therefore, acts as an emission promoter and in conventional MECA is the means by which the sample is introduced into the flame.

The cavity is pitched at an angle of about 7–10° downwards from the horizontal and a few millimetres (usually 10–20 mm) above the burner head. The size and shape of a cavity should be based on the size, composition and shape of the flame, and thus its design is closely related to the burner system. A typical cavity is 8 mm in diameter and 5 mm deep, but many other cavities differing in dimensions have been proposed for specific applications (4). The composition of the probe and especially the cavity wall is particularly important (See Chapter 2) and is determined by the specific application. The most widely used materials for cavity construction are aluminium or stainless-steel, but many other materials have also been used.

MECA cavities have been classified into two main categories as follows (5):

(a) *Cavities that heat up continuously after introduction into the flame.* The first cavity used in conventional MECA was a stainless-steel Allen screw which had an hexagonal aperture (maximum width 15 mm, volume 45 μL) and a 30 mm long stem ((Figure 3.4 (a)) (6). In a modified design, the cavity (4 mm diameter × 4 mm depth) was screwed into a base which fitted into the cavity holder ((Figure 3.4

INSTRUMENTATION 47

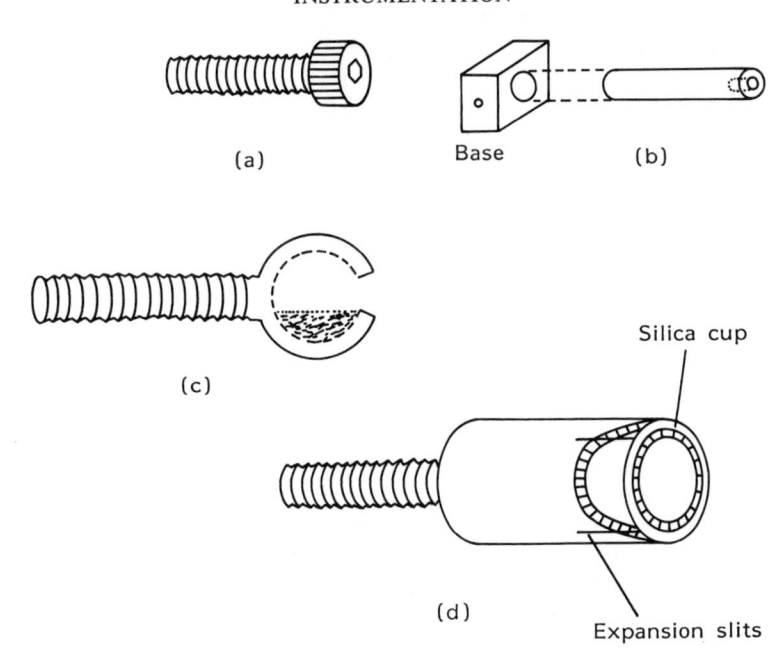

Figure 3.4. Cross sections of various designs of cavities for conventional MECA.

(b)) (7). The opening of the cavity can be modified ((Figure 3.4. (c)) for the deposition of inert solids impregnated with analyte, e.g. silica gel with adsorbed sulphur dioxide from atmospheric air (8).

Aluminium, copper and carbon can also be used in some applications. For example, the emission from a volatile sulphur compound will be more intense when injected into a cavity made of aluminium than into one made of stainless-steel since the former will conduct away heat faster than the latter and will remain cooler during analysis (Salet phenomenon, see Chapter 2). The formation of carbon monoxide within a carbon cavity continuously renews the inner surface and creates a reductive environment which may be conducive to particular kinds of emission.

Formation of refractory compounds (solid residue) within the cavity is a problem which often appears in conventional MECA. The cavity must then be replaced because its characteristics (reflectivity, size) change and the emission intensity is altered. The effect can be overcome by using a carbon cavity or a silica-lined cavity in which a silica cup is fitted into a stainless-steel cavity or a copper support ((Figure 3.4. (d)) (9). The cup is washed between injections and the solid residue is removed.

The use of tantalum cavities (10) is not satisfactory because a white film of tantalum(V) oxide forms. The film is not adherent, falls out, and the cavity size increases with time. Titanium and zirconium cannot be used because a

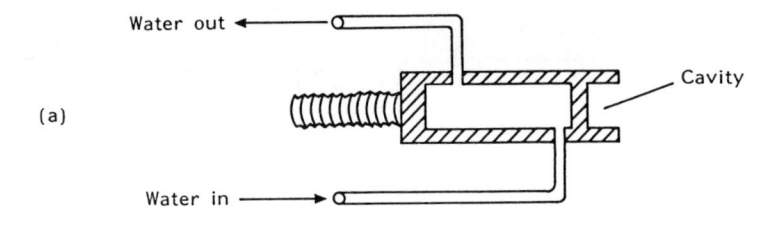

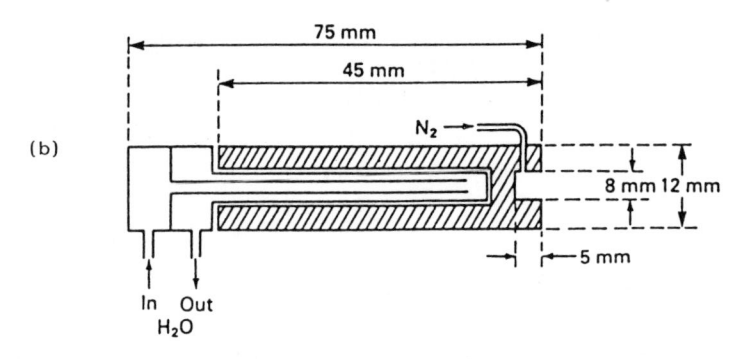

Figure 3.5. Cross sections of various designs of water-cooled MECA cavities

strong incandescence appears immediately after the cavity is introduced into the flame.

Very often cavities should be coated internally with a suitable material in order to stimulate emissions which otherwise do not occur. Thus, an indium-lined cavity is used for the generation of the indium(I) halide emissions from halide-containing samples (11). This is achieved by initially heating a piece of metallic indium in a copper-lined cavity until a bright red indium emission is observed and maintained for a few minutes. Excess molten indium is then poured out of the cavity. The bright layer of metallic indium coats the copper, presumably by forming an alloy. A modification, where several layers of a fine copper mesh are embedded in a carbon cavity before treating with indium metal, gives a cavity with prolonged life (> 100 analyses). Copper (12) and tin (13) can be used in a similar way. It is clear that metallic cavities cannot be used when strongly acidic halide solutions are injected because the corresponding metal halide emissions are generated. In this case, a carbon cavity internally coated with pyrolitic carbon, a silica-lined cavity, or an aluminium cavity should be used.

(b) *Cavities that maintain constant temperature when heated in the flame.*

INSTRUMENTATION

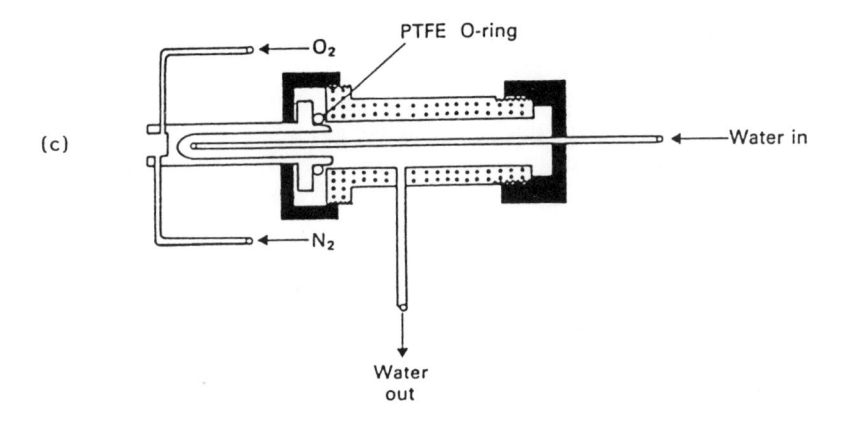

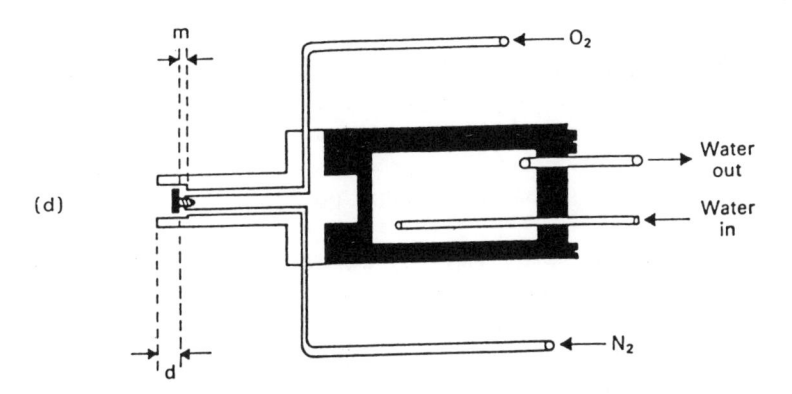

Figure 3.5. (*continued*)

Constant temperature inside the cavity is achieved by passing a continuous flow of tap water through the cavity body ((Figure 3.5 (a, b)). This cavity is rarely used in conventional MECA but it is almost exclusively used in connection with vaporization systems. The vapours generated are introduced into the cavity via a side duct using an inert gas for purging.

Various designs of water-cooled cavities exist (Figure 3.5) but the design in Figure 3.5 (d) has some distinct features. The cavity is made of aluminium and is mounted on a brass cooling chamber. Nitrogen and oxygen, if required, are introduced into the cavity from the rear through carefully drilled passages. The head of a screw separates the cavity space into the mixing area and the actual cavity. The incoming gases mix behind the screw head and are then transported around its edges into the cavity. The cavity can be removed from the cooling block without turning off the cooling water. Furthermore, mixing of the incoming gases prior to introduction into the actual cavity improves the signal-to-noise ratio (14).

Another cavity design is the *cavity with gas supply*. This design promotes molecular emissions by introducing a reactive gas into the cavity. In this case, the gaseous inner environment of the cavity is altered so that emissions which would not otherwise occur are allowed. The two main types of cavity with gas supply are the *oxy-cavity* and the *hydrogen chloride cavity*.

The oxy-cavity ((Figure 3.5 (c)) is a cavity into which a slow flow of pure oxygen is supplied through a stainless-steel tube (15). Oxygen promotes molecular emissions from elements such as arsenic, antimony, tin, boron and silicon (see Table 2.1) either through its presence in the excited emitting molecule or radical (such as BO_2), or through oxygen atom combination reactions, such as:

$$AsO + O \longrightarrow AsO_2 + h\nu$$

When the oxy-cavity is supplied with oxygen, S_2 and Se_2 emitting species are destroyed, PO rather than HPO is generated, and metal halide emissions are replaced by metal oxide or metal hydroxide emissions.

The oxygen supplied to the cavity burns with hydrogen supplied by the flame to form a small hydrogen–oxygen flame within the cavity. In this situation a pronounced OH emission is observed. The conditions within the cavity resemble those at the edge of the flame and, hence, the hot environment encourages atomic emissions, such as those from lithium, potassium, sodium, and thallium (16).

The hydrogen chloride cavity is a cavity into which pure hydrogen chloride is supplied for stimulation of metal halide emissions (17). The reactivity and toxicity of hydrogen chloride cause many problems which must be overcome before the potential applications of this cavity are fully investigated.

Another cavity which must be classified separately from those already mentioned is the flame-containing cavity, which has been used extensively by Bogdanski *et al.* (18) for the determination of boron, selenium and other metalloids. The cavity has three side ducts. Hydrogen and nitrogen are introduced into the cavity through two of the ducts while the third is used for the introduction of analyte vapours which are carried by nitrogen. In this design, the flame is maintained within the cavity (Figure 3.6). The cavity is also water-cooled to ensure a constant and relatively low temperature. Since the flame is retained within the cavity, external heating and careful positioning of the cavity in the flame is unnecessary (see Chapter 2). Although the philosophy of this MECA variation is the same as others, i.e. formation and stabilization of the excited molecules within the cavity, it is significantly different.

3.2.2.2 Cavity Holder

The conventional MECA system must include a device for repeatedly intro-

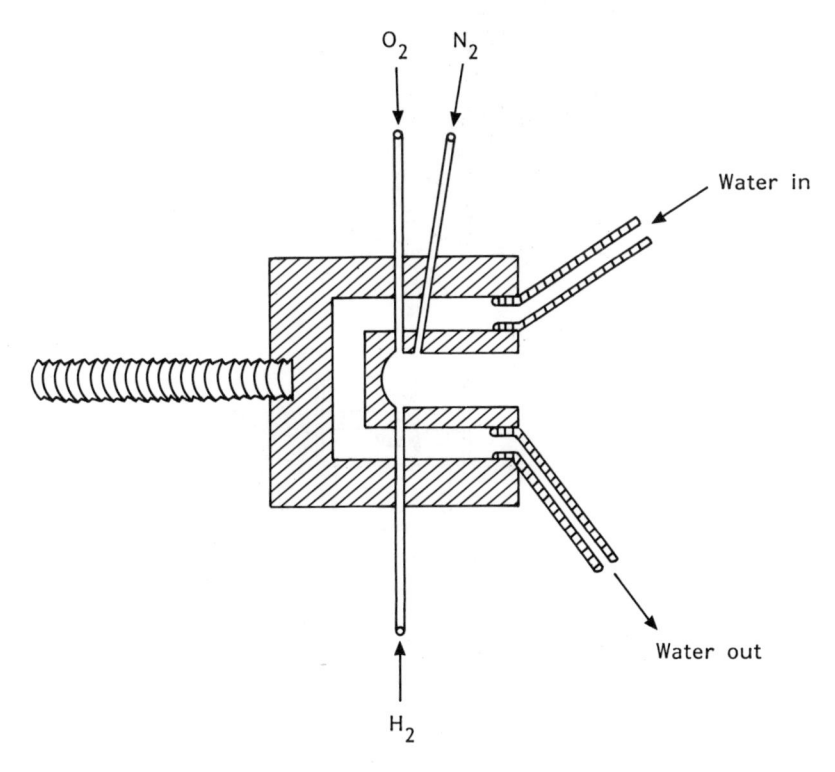

Figure 3.6. Typical flame containing MECA cavity.

ducing the cavity into the flame, and subsequently removing it to cool when the measurement is complete so that the next sample can be introduced. Such devices have usually involved rotation of the probe through 90° into a predetermined position in the flame (9), since in this configuration liquid samples can be injected into the probe while it is held almost vertically (Figure 2.5). This cavity holding apparatus is used in conventional MECA.

A different cavity holder was constructed by Wanegen and Fernando (2). This holder was designed to provide continuous spinning of the cavity probe (rod) using a gear assembly and to thus spread the sample solution into a thin film over the entire inside surface of the cavity during the measurement (Figure 3.7). The cavity used was a cylindrical silica cup fitted into the end of the rod. The rod was cooled using a copper cooling coil wrapped around it.

Wanegen and Fernando (2) also used a modified burner consisting of a copper tube about 12 mm in diameter. The tube was sealed at one end and a 1.5 mm hole was drilled through the sealed end. A cool jet-shaped flame was produced by passing a mixture of hydrogen and nitrogen through the tube.

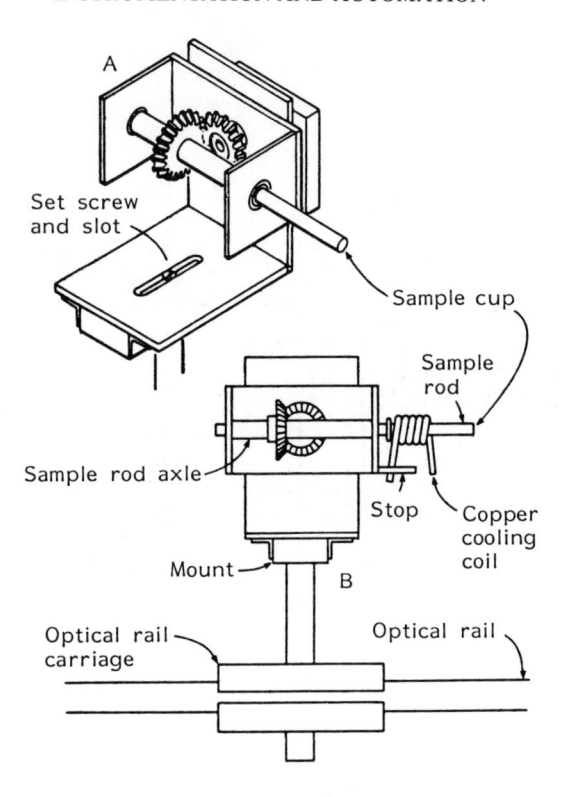

Figure 3.7. Modified rotatable MECA cavity holder. (A) View of gear assembly and mount for spinning sample cup. (B) Side view of gear assembly and mount for spinning cup, showing the position of the cooling coil and optical rail. Reprinted with permission from Wanegen and Fernando, *Anal. Chem.*, **57**, 2743. Copyright (1985) American Chemical Society.

Combining this specially designed holder with the modified burner, the authors reported increased reproducibility and better resolution of the recorded emission peaks. More specifically, with this system they were able to determine sulphur as sulphate in the range 6–640 μg mL^{-1} (by measuring peak areas) with a reproducibility of 1–2%. This level of reproducibility was much better than that previously reported by other researchers using conventional burners. Furthermore, a complete resolution of the emission peaks of sulphite and sulphate was observed (Figure 3.8).

For gas generation detection, a device for permanently holding the cavity in the desired position in the flame must be employed, and has been used by Burguera *et al.* (19). By using fine verniers, these investigators were able to adjust the cavity angle, and vertical and horizontal cavity positions very precisely so that optimal emission intensities were obtained (Figure 3.9).

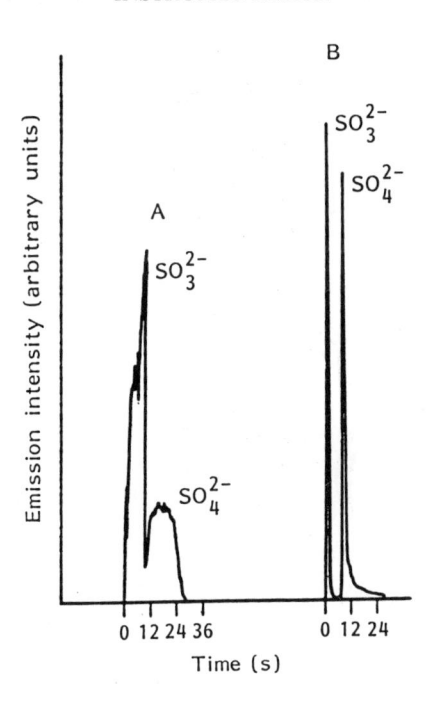

Figure 3.8. S_2 emission-time profiles: (A) conventional sample cup and burner and (B) spinning sample cup and angled jet flame. Reprinted with permission from Wanegen and Fernando, *Anal. Chem.*, **57**, 2743. Copyright (1985) American Chemical Society.

3.2.3 Optical Unit – Readout System

The major parts of the optical unit are the entrance slit, monochromator, the exit slit and the light detector (Figure 3.1). Since most molecular emissions are bands, a high resolution monochromator is not necessary. A conventional photomultiplier tube (PMT) is used in line with the cavity source, detecting the light emitted. Extended amplification is possible due to the very low background emission of the hydrogen–nitrogen flame (see Chapter 2).

In most applications, the absence of spectral overlap between the analyte emission and that of any other species allows the use of a simple optical cut-off filter instead of a monochromator. This modification increases sensitivity, since a wider spectral range of the generated emission is collected, and losses of radiation due to the length of the optical path are minimized.

A photometric readout and a chart recorder are the major components of the readout unit. The recorder should be comparatively fast, since certain conventional MECA responses appear within 0.2–0.3 s after cavity introduction into the flame. If necessary, a damping circuit can be inserted between the photometric readout and the recorder. For details about the design, operational features, and characteristics of the components of a

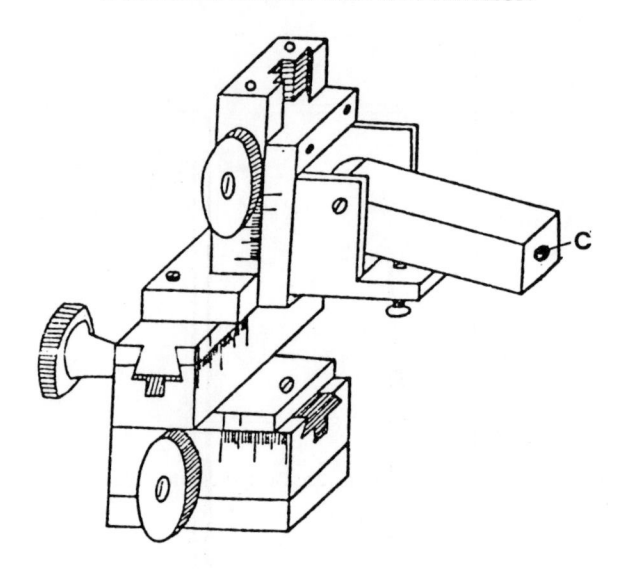

Figure 3.9. Adjustable cavity holder for MECA–VAP experiments. The cavity is screwed into hole C. Reproduced by permission of Elsevier Science Publishers from Ref. 9.

common spectrophotometer, the reader can refer to a variety of dedicated books.

Calokerinos and Hadjiioannou (1) have made an extensive study of how various experimental parameters affect the S_2 calibration by using a spectrometer containing items shown in Table 3.1.

3.2.4 Gas Generation Systems

The gas generation system is a special accessory to MECA–VAP instruments

Table 3.1. Items of a Typical MECA Spectrometric Unit

Item	Model	Characteristics
Monochromator	Heath EU-700	0.35 m Czerny–Turner optical system with grating (1180 grooves mm^{-1}) blazed at 250 nm; dispersion 2 nm mm^{-1}
Photomultiplier	EMI 6256B	End-window type, S11 response, fused silica window
Power supply	Kepko	0–1000 V, 0–20 mA
Photometric readout	Heath EU-703–31	Maximum output 0.1 V at each current range
Recorder	Sargent–Welch XKR	0.47 s full scale deflection

Reproduced by permission of Elsevier Science Publishers from Ref. 1.

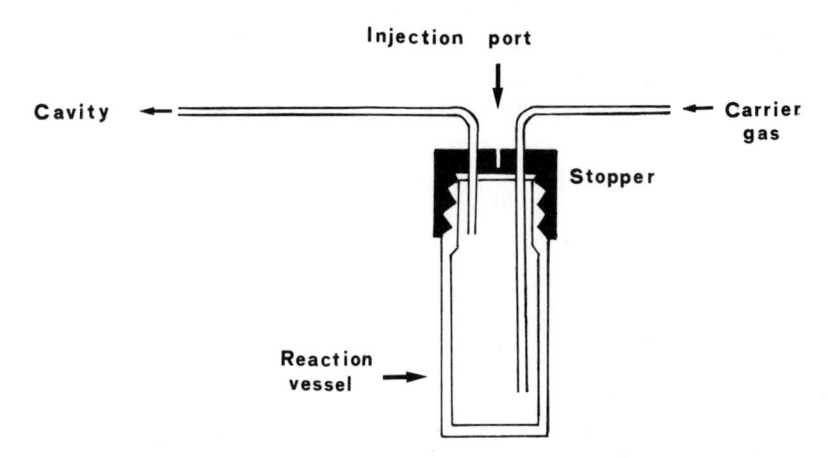

Figure 3.10. Typical volatilization system used for the generation of a gaseous product from the analyte.

and is used when the analyte has to be introduced into the cavity as a vapour. A typical gas generation system consists of a glass reaction vessel, the top of which is closed with a PTFE stopper (Figure 3.10). The reaction vessel contains the necessary reagent, which reacts with the sample to produce the gaseous product. The stopper is fitted with two holes, one for the carrier gas inlet, and the other for the the analyte gas outlet. A septum, which serves as an injection port for the introduction of an aliquot of sample solution, is also incorporated into the stopper. A glass syringe is used for injecting the sample into the reaction vessel. The gaseous product which is generated in the reaction vessel is transferred by an inert gas through stainless-steel tubing to the cavity. The glass vessel may be heated and then held at a constant temperature in order to increase the quantity of gaseous product evolved and to accelerate the evolution rate of the gas mixture from the solution of reactants. In such cases, care must be taken to avoid purging water vapours into the cavity. Gas generation devices with several modifications and/or improvements have been used extensively in MECA–VAP applications (19–22).

3.3 COMMERCIAL INSTRUMENTS

3.3.1 Conventional MECA

3.3.1.1 MECA-22 Spectrometer

This instrument was the first commercially-available MECA spectrometer (Anacon Inc., Houston, TX, USA) designed exclusively for conventional applications. It employs a pre-mixed nitrogen–hydrogen (air) flame with a burner adjustable in one vertical and both horizontal directions. Each gas is introduced into the burner after passing through a flow meter and mixing chamber.

The cavity holder is a rotatable cylinder (as shown in Figure 2.5). The sample solution is injected into the cavity when it is vertical, after which the cylinder is rotated through 90° to position the cavity reproducibly in the flame. When the emission responses from a given sample have been recorded, the cylinder is returned to its initial position and the cavity probe, now back to the original injection position, is allowed to cool before injection of the next sample. Cooling is usually accomplished with an air blower.

The inclination of the cavity aperture to the flame axis can be altered to give optimal emission intensities by making appropriate adjustments to the cylinder. The bottom edge of the cavity is about 10 mm above the top of the burner.

The emission from the cavity is passed through an 8.5 nm slit and a grating monochromator (590 grooves/mm blazed at 300 nm) to a photomultiplier (RCA 1P28), the output of which is fed to a potentiometric recorder (e.g. Oxford 3000 or Leeds and Northrup Speedomax XL). The instrument also incorporates a built-in electronic integrator that allows integration of the peak areas.

Using the above described instrument and a removable carbon cavity, Flanagan and Downie (23) determined sulphate at trace levels in high purity water. The procedure followed included a preconcentration step where successive aliquots of the liquid sample were placed in the cavity and then partially evaporated by heating on a hot plate. The last aliquot was allowed to evaporate completely and the cavity was removed from the plate and allowed to cool. Ten microlitres of 0.5% (v/v) orthophosphoric acid were then injected into the cavity which was inserted into the instrument, rotated into the flame and the emission recorded. By using this procedure, sulphate in the range 20–300 ng mL^{-1} was determined within 10 minutes.

3.3.2 Gas Generation Detection

3.3.2.1 MECA–VAP

During the development of conventional MECA, the advantages of a gas generation system became apparent. However, the MECA-22 instrument was not readily adapted to this methodology, so a new instrument, named MECA–VAP, was designed. This instrument, when eventually modified for commercial production by Doma Instruments, included a digestion/vaporization unit (DIVAP-201) and a detector unit (MEP-101), and was versatile enough for both conventional MECA and gas generation experiments, as well as accommodating possible future advances of the technique.

3.3.2.1.1 Design and Characteristics of MECA–VAP. The MECA–VAP instrument consists of two separate units, one for the cavity and the other for the electronic components. The cavity unit houses the cavity, the burner, the photomultiplier tube and the flame gas plumbing. The cavity is divided into two compartments by a screw and is mounted on a cooling chamber. The cavity projects downwards, usually at a fixed angle of 7° to the horizontal but it can be moved in all three co-ordinates. Cooling water, nitrogen and/or oxygen supplied to the cavity are controlled by individual flow meters.

The burner is the same as in the MECA-22 spectrometer. It is mounted on a suitable holder and can be raised or lowered. The flame gases enter a mixing chamber and proceed to the burner. These gases are controlled by solenoid valves and are ignited by an electrical spark igniter which stands just on top of the burner head. The igniter operates automatically 3 s after turning the flame gases on, to allow them to flow out of the burner head. A thermocouple positioned 11 cm above the burner head serves as a safety device, shutting off the flame gas solenoid when a sufficient temperature decrease is sensed. The thermocouple is activated when the flame is accidentally extinguished for any reason.

The PMT is mounted on a holder secured to the optical bench on the bottom of the instrument. The detector is held by screws which can be adjusted for alignment with flame and cavity.

The gas generation system consists of an 8 mL glass bottle, serving as reaction vessel, which screws into a Rulon nut secured on a stainless-steel cap mounted on a small brass table. The table can slide in or out according to the need of the operator. Liquid samples are injected through a silicone septum into the glass reaction bottle. The carrier gas inlet tube extends to close to the bottom of the bottle and the vapours are transferred to the cavity through an outlet tube, maintained at 150° by electric heating. The nitrogen purge gas for the vaporization system is supplied directly from the flame nitrogen line. A pair of electrically-controlled solenoid valves direct the carrier gas to the cavity via the vaporization system or via a by-pass route. A micrometer valve is able to adjust the resistance of the by-pass line to match that of the sample line. In the by-pass mode, the vial may be removed from the holder without any change in the background emission.

The power supply includes a 'MAINS' switch, the 'LINE HEATER' which heats the vapour duct between the reaction chamber and the cavity, and a voltage adjuster for the PMT. The gas control panel controls the solenoid switches of the flame and the gas generation system. It also contains the main switch for the flame gases and the 'IGNITE' button, which activates the ignition spark and must be held down manually until the flame is ignited. Flame ignition is indicated by the 'LIT' light coming on while the 'FLAME OUT' light goes off. The 'WATER' light indicates that water is flowing through the cavity.

Three button settings are available for controlling the nitrogen carrier gas. In the 'SAMPLE LINE CLOSED' position, carrier gas is completely shut off from the cavity. In the 'BY-PASS' position, nitrogen is introduced into the cavity via the by-pass route, while in the 'SAMPLE LINE' position the gas is diverted through the reaction chamber. Indicator lights show which function is in operation. The amplification module contains settings for sensitivity, 'BACK OFF' (background), and 'DAMPING' control. The 'TRANSMISSION' meter indicates the level of emission response. The readout and integration module contains controls for integration. Emission responses can be recorded either directly or in the integrated mode by connecting a recorder to the 'REC' terminals.

3.3.2.1.2 Design and Characteristics of the MEP-101/DIVAP-201. The instrument consists of a detector unit (MEP-101) and a digestion/vaporization unit (DIVAP201). The MEP-101 unit houses the cavity, burner, PMT and flame gas plumbing.

An aluminium cavity is mounted on a water-cooled chamber and is surrounded by a ceramic insulator that maintains the cavity at a low enough temperature for the Salet phenomenon to be effective, but high enough to prevent water condensation on its outer surface. A thermocouple end, protruding from the insulator, is situated in the flame. If the flame is accidentally extinguished, the thermocouple cools down and switches off the hydrogen and nitrogen supply. Another thermocouple monitors the cavity temperature. If the water flow is disturbed, the thermocouple warms up and automatically turns the flame gases off to avoid overheating and possible destruction of the cavity and cooling chamber. The burner is made of stainless-steel and gases are supplied premixed to the flame. Flow gas rates are adjusted by flow meters housed in the unit.

Emissions pass through a blue filter (λ_{max} = 384 nm) onto the PMT (RCA 1P28). The PMT window is covered by a shutter which is on the same horizontal axis as the cavity. Thus, unnecessary flame radiation is excluded from the PMT and the signal-to-noise ratio of the instrument is greatly improved. At the rear of the cavity unit, a continuously operating fan circulates fresh air and prevents the photomultiplier from warming up after prolonged use. If the flame is accidentally extinguished and before the gas solenoids are automatically switched off, the air flow removes hydrogen from the unit and eliminates any potential hazard during reignition.

The electronic unit controls signal amplification through coarse, damping, and background dials. The output from the PMT can be fed to a potentiometric recorder.

The MEP-101 unit has been designed primarily for the analysis of gaseous samples, especially sulphur-containing compounds. In this instance, the intensity, I, of the S_2 emission generated is related to the concentration, C, of the analyte in the sample by the following equation:

$$I = KC^n$$

where K and n are constants (see Chapter 2). The electronic part of the instrument incorporates a function generator which linearizes the calibration curve by calculating and displaying the nth root of the emission intensity. Thus, n is the slope of the log–log straight line graph of a given compound. The value of n is used to express the sensitivity of the determination. Samples are delivered to the cavity as gases, either because this is how they are originally prepared, or because they are converted into volatile products by chemical reaction or by being passed through a gas chromatography.

The digestion/vaporization reaction which is necessary to convert the sample into gaseous products is carried out in the DIVAP-201 using 10 mL glass vials. The vial is screwed into a gas-tight PTFE top that is held by springs to eliminate breakage. The sample solution is injected into the vial through an injection port containing a silicone septum. Solid samples can be transferred into the vial prior to securing it in position. If necessary, the vial can be heated at an adjustable rate by a suitable heating device. Vapours generated in the vial are swept into the cavity by a flow of nitrogen after passing through a water cooled condenser.

The combination of the DIVAP-201/MEP-101 has been used successfully by Calokerinos and Townshend for the determination of sulphur dioxide in air. The MEP-101 was also used as a gas chromatographic detector, in conjunction with a suitable gas chromatograph (24). The simplicity, sensitivity and safety features of the MEP-101 make it attractive as an instrument that can be used by unskilled operators with excellent results.

3.3.3 Gas and High Performance Liquid Chromatographic Detection

Based on general considerations, it is apparent that MECA can provide the basis for sensitive and selective determinations of a wide variety of elements, especially non-metals. The ability of the cavity to detect trace quantities of volatile compounds makes it attractive as a detector for gas chromatography (GC). Several successful applications of MECA to GC are provided below.

The essential components of a combined GC–MECA system are shown in Figure 3.11. The outlet port of the GC column is connected to the MECA cavity by a short stainless-steel transfer tube which is maintained at the same temperature as the GC oven by wrapping a heating tape around it. This apparatus was used by Burguera et al. (25) for the separation and detection of chlorinated solvents and pesticides. They combined a Varian AA–275 atomic absorption spectrophotometer operated in the emission mode with a Perkin–Elmer F11 gas chromatograph. The spectrophotometer was equipped with a cavity support (similar to that illustrated in Figure 3.10) bearing an indium-lined cavity. Emission responses were recorded on a Varian 9176 strip chart

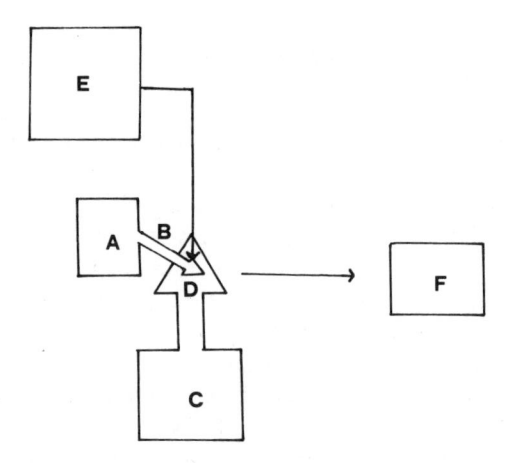

Figure 3.11. Schematic diagram of a MECA–GC system (not to scale). A: cavity holder; B: probe and cavity; C: burner; D: flame; E: gas chromatograph; F: detector.

recorder. The chromatograph was fitted with a stainless-steel column (4.0 mm o.d., 2.0 mm i.d., 200 cm in length) packed with 3% OV–101 on acid-washed, HMDS-treated Chromosorb W, 80–100 mesh. The authors reported a linear response of the detector up to 60 μg mL^{-1} of chlorine. The precision was 2.4% and the limit of detection was 1.2 ng of chlorine (see Chapter 7).

Burguera and Burguera extended this application to the separation and detection of bromo and iodo compounds, using the same instrumentation with a water-cooled cavity support. This support was made of a brass rod and was placed around the cavity. A Varian BD atomic absorption spectrophotometer was used in this case (26).

Henden *et al.* used an oxy-cavity (27) and a flame-containing cavity (28) for the detection of arsenic, germanium, tin and antimony compounds after gas chromatographic separation.

The cavity has also been operated as a detector in liquid chromatography (LC) by Cope and Townshend (29). This detector consists of a Duralumin disc (10 cm in diameter) with forty cavities drilled into its circumference. The disc is powered by a stepper motor and rotated in an anticlockwise direction as shown in Figure 3.12. The motor is controlled by a digital indexer (Unimatic Engineers Ltd, London, UK) and a custom-built timer unit. One drop of effluent is collected per cavity. The cavity then moves on and the following drop is collected in the next cavity and so on. The time between steps is set so that the cavity disc is synchronized with the drop time of the effluent. The delay must also be sufficient to allow the emission to be generated within the cavity as long as it remains in the flame. Usually, three seconds are enough.

The surface temperature of the disc is maintained constant by circulating cooling water through an inner brass cooling chamber. Solvent evaporation

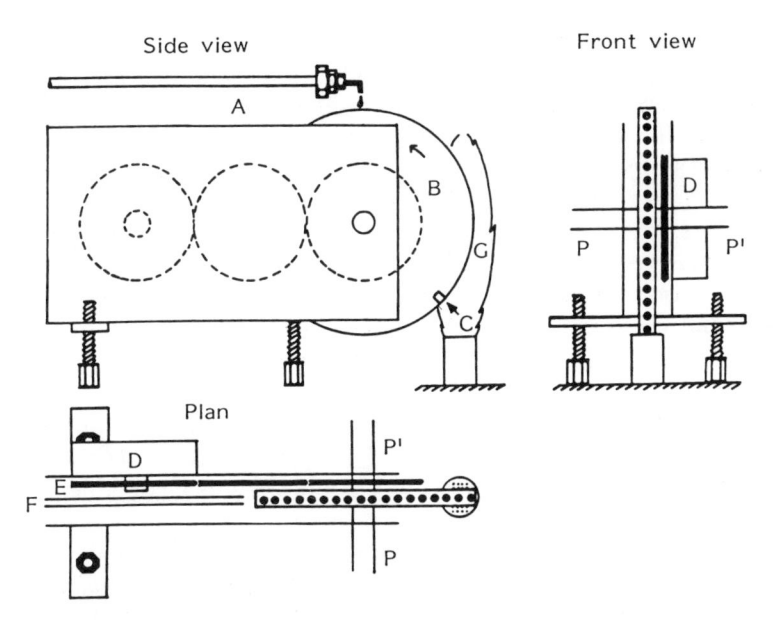

Figure 3.12. MECA–LC detector. (A) column; (B) duralumin disc with 40 cylindrical cavities drilled into its circumference; (C) cavity position for measuring emission; (D) stepper motor; (E) gears connecting stepper motor to disc; (F) tube form air pump to aid in solvent evaporation; (G) flame; (P,P') inlet and outlet pipes for water. Reproduced by permission of Elsevier Science Publishers from Ref. 29.

takes place by adjusting the temperature of the disc to an appropriate value and by using a small air inlet. After solvent evaporation, the sample band, which is often distributed between about seven cavities (see Chapter 6) is carried round into the flame. Emission intensities reach their maximum values when the cavity enters the flame edge at 45° to the horizontal. After the emission intensity has been recorded, the cavity passes through the bulk of the flame and steps around to the vertical position where it collects more eluent.

In this study, the cavity holder was placed within a modified Evans Electroselenium 240 atomic absorption spectrophotometer operated in the emission mode. An Altex model 100 dual position reciprocating pump was used for pumping the solvents, while an Altex injection valve was used with a 20 μL sample loop. The signals were recorded on a Servoscribe 1S chart recorder. Organic compounds emerging from both normal and reversed-phase columns were detected with a reproducibility of about 3%. Results were obtained by using peak area measurements.

3.4 AUTOMATION

The reproducibility of conventional MECA is limited, especially for fast-emitting species where the control of experimental variables is of paramount importance. These variables include parameters such as the position and the residence time of the cavity within the flame, the initial temperature of the cavity (dependent on the cooling time before the next sample injection), the time allowed for solvent evaporation before cavity introduction, and the rate at which the cavity is introduced into the flame. Furthermore, the whole procedure is usually cumbersome, time-consuming, and dependent on the individual skill of the investigator. In MECA determinations based on the gas generation technique, precise control of basic experimental variables is inherently easier, but the requirements of excellent reproducibility, rapid and sensitive measurements, and uncomplicated analytical steps still remain.

The search for improved precision and easier operation led to the design and construction of automated MECA analysers (30,31). These developments, which have been fully justified by experiment, took place primarily in the 1980s.

According to the IUPAC recommendation, automation is defined as the '*use of the combination of mechanical and instrumental devices to replace, refine, extend or supplement human effort in the performance of an analytical process, in which at least one major operation is controlled without human intervention, by a feedback mechanism*' (32). The use of mechanical devices just to replace, refine, extend or supplement human effort is defined as mechanization. Even though the distinction between these terms is clear, in many cases the term automation has been improperly used to describe any development which minimizes human intervention in chemical analyses.

Almost any repetitive analytical determination can be automated. Automation may be applied to any or all the steps common to most analyses, for instance (1) sampling, (2) separation of analyte from matrix material, (3) addition of one or more reagents and subsequent mixing, (4) measurement, including standardisation and calibration, and (5) collection, manipulation, presentation, storage and archiving of the analytical data. In a fully automated method, a microcomputer is necessary to program the entire sequence of events and to respond to information fed back from the measuring system.

In comparison with manual analysis, an automated analytical system offers advantages such as better analytical precision, easier operation, and convenient handling of toxic or unstable reagents. More samples can be processed in the same time without results being affected by human error, fatigue, or by momentary inattentiveness. Furthermore, the economic advantages of automation are obvious.

Attempts by researchers to automate MECA follow. However, it must be stressed that some of these systems do not fall strictly within the IUPAC definition of automation (33).

3.4.1 Conventional Automated Analysers

The schematic diagram of the first automated MECA analyser is shown in Figure 3.13. The instrument incorporates automatic sample dispensing into the cavity and mechanical insertion of the cavity into the flame. Both operations are microprocessor controlled. The instrument consists of a moveable cavity device (MCD), a Varian ASD-53 automatic sample injector with a control unit and rotating sample carousel, a Pye Unicam flame emission spectrometer equipped with an EMI-S11 PMT, and an Oxford 3000 fast response recorder.

The MCD introduces the cavity into the flame and withdraws it automatically. It consists of a synchronous motor that is connected by a flexible joint to a gearbox. The gearbox is attached to an egg-shaped cam bearing four micro switches on its periphery. The micro switches are connected to a microprocessor-controlled timer. Two of these switches are used to control the time that the cavity remains in or out of the flame and are programmed via the microprocessor. The third switch is used to increase the cooling time of the cavity when required, and the fourth synchronizes the performance of the automatic sample injector with the MCD.

The microprocessor-controlled timer is used to schedule the cavity movement time cycle. It consists of an INS 8060 N monolithic, 8-bit, N-channel microprocessor with an address capability of 64 K. It has also a direct memory access, two sense inputs, bus logic and a serial input–output port.

This analyser was used by El-Hag and Townshend (31) for the determination of phosphorus anions in inorganic samples after an ion-exchange batch procedure. They used a 4×4 mm carbon cavity cut at the end of a 37 mm long carbon rod. The rod was housed in an aluminium holder attached to the MCD.

The analytical procedure, carried out under optimized experimental conditions (see Table 3.2) proceeds as follows. The pre-treated sample solution is transferred into a vial in the carousel unit. At the beginning of the analysis of the first sample, the microprocessor unit sends a pulse signal via the MCD to the control unit of the ASD-53 to begin the injection of the samples into the

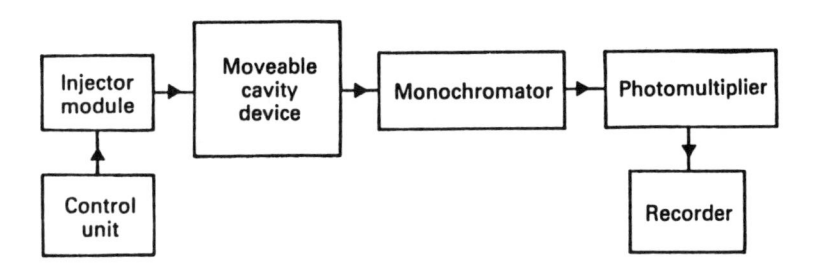

Figure 3.13. Blocked diagram of the automated molecular emission cavity analyser (not to scale). Reproduced by permission of the Royal Society of Chemistry from Ref. 30.

Table 3.2. Optimum Conditions Used for the Determination of Inorganic Phosphorus Compounds by Automated MECA

Condition	Optimum value
H_2 flow rate	2.9 L min^{-1} [1]
N_2 flow rate	5.0 L min^{-1}
Air flow rate	5.5 L min^{-1}
Distance of cavity front protruding inside flame[1]	6 mm
Cavity centre to burner top distance	4 mm
Residence time of cavity in flame	10 s
Cooling time of cavity	90 s
Wavelength	528 nm
Slit width	0.3 nm

[1] 1–2 mm in the presence of sulphate. Reproduced by permission of the Royal Society of Chemistry from Ref. 33.

cavity. Once the sample has been injected, the cavity moves to a pre-selected zone of the flame by the action of the MCD.

After the emission intensity has been recorded, the cavity is returned to its initial position for cooling before the next sample is injected. When the last sample has been analysed, the injector unit sets itself at the standby position. Using this procedure, relative standard deviations for phosphorus determinations (see Chapter 6) were improved from 4.5% (manual procedure) to 0.9%, and it was possible to analyse at least 30 samples per hour.

Using the same analyser but different experimental conditions, Evmiridis and Townshend determined the sulphur-containing compounds thiourea and promethazine (34). The procedure was similar. They accomplished much greater sensitivity and calibration linearity by including a solvent evaporation step prior to MECA determination. Using this technique, they were able to simultaneously determine thiocyanate and sulphate in binary mixture.

The construction of the automated analyser for conventional MECA measurements described above allows precise control of the cavity's movements, by programming flame residence and cooling times. The microprocessor also controls sample injection into the cavity and, when necessary, can be used to control the level of solvent evaporation. This instrument makes it possible to undertake a series of determinations without human intervention, and offers a great improvement in analytical precision. The method also has the advantages of good throughput, easy operation and convenient use.

Burguera and Burguera have designed and constructed an automatic instrument for MECA measurements in the conventional mode by using a flow injection (FI) system for delivering the liquid samples to the cavity (35–37). Flow injection systems such as this have also been proposed as a means of interfacing

AUTOMATION

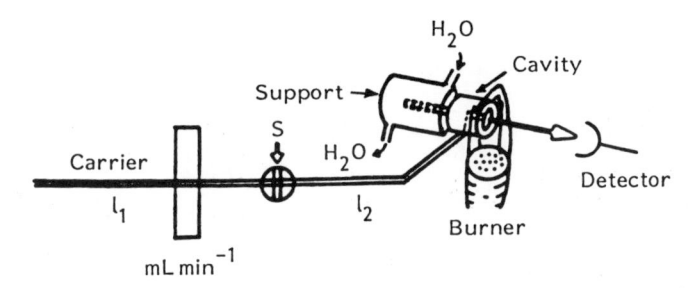

Figure 3.14. Schematic diagram of the FI–MECA system. (S) rotary sample injector, (l_1) Teflon tubing, (l_2) stainless-steel tubing. Reproduced by permission of Elsevier Science Publishers from Ref. 35.

a sample tray with the nebulizer of an AAS instrument in the well-known FI–AAS technique (38).

This FI–MECA system is shown in Figure 3.14. Burguera and Burguera used a Varian (model AA–1475) atomic absorption spectrophotometer operated in the emission mode and equipped with a sample holder support device and a circular emission burner. A water-cooled steel MECA cavity (5 mm diameter, 8 mm deep) situated continuously in the flame was used. The flow injection system was connected to the MECA cavity by a steel tube (0.5 mm i.d.) screwed into a hole drilled on its side wall.

The flow injection unit, used to transport the sample to the cavity, consisted of a Sage pump (model 375-A) with adjustable flow control and a Rheodyne (model 7125) rotary valve. Tubing of 0.5 mm i.d. was used. Maximum sensitivity and reproducibility were obtained by careful control of flame composition and flow rate of sample, both of which determined the rate at which solvent evaporated. Using this procedure, Burguera and Burguera were able to make successful determinations of sulphur and phosphorus in a variety of compounds.

3.4.2 Automated Gas Generation Analysers

Continuous flow (CF) and flow injection are two widely known techniques in which samples are allowed to react with reagents in a continuously flowing system. In the CF technique, the sample is injected into the flowing reagent stream where mixing of individual samples (carry over) is eliminated by the introduction of air bubbles (segmentation).

In the FI technique, the sample is injected into an unsegmented continuous stream of a liquid solvent which is then mixed with the reagent stream by diffusion. Analysis by CF is preferred for reactions in which the determination takes longer than 2 minutes and where three or more reagents are required. The technique of FI is usually used in determinations that can be completed in less than

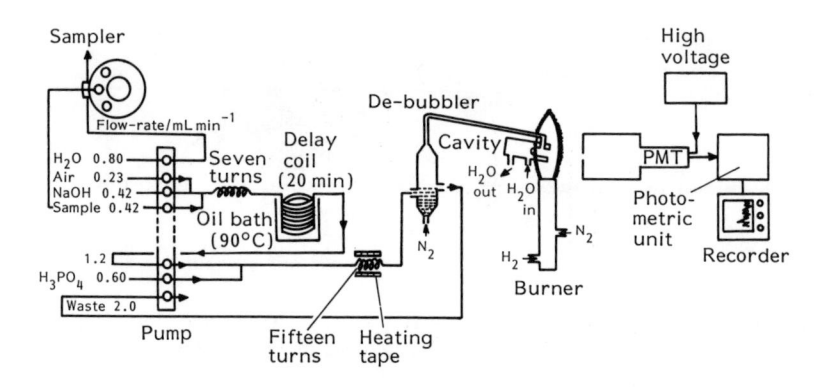

Figure 3.15. Schematic diagram of the continuous flow molecular emission cavity analyser (not to scale). Reproduced by permission of the Royal Society of Chemistry from Ref. 42.

30 s and where only one or two reagents are used. These two techniques are complementary to each other. Both have been used successfully in conjunction with the MECA cavity, which in this case acts as the detector for gases produced within a flowing stream.

The continuous flow molecular emission cavity analyser (CF–MECA) was initially developed for the detection of sulphur dioxide and hydrogen sulphide and therefore for the indirect determination of sulphur compounds that produce these gases by chemical reaction (39–43). The schematic diagram of the CF–MECA instrument used for the determination of thiamine and cephalosporins in pharmaceuticals (42,43) is shown in Figure 3.15. It consists of a Technicon proportionating pump and an auto-sampler with a capacity of 40 samples (40 sampler II, Hook and Tucker Instruments Ltd). The proportionating pump delivers sodium hydroxide and the sample through tubing to a mixing coil and then to a delay coil, where alkaline hydrolysis of the sulphur compound to sulphide occurs. Air is introduced via a third tube.

The delay coil outlet is re-connected to the pump so that undesirable pulsing is avoided. A solution of orthophosphoric acid is then added to the flowing stream, which is transported to a second mixing coil heated at 45°C. Reagent flow rates and, therefore, their relative proportions are determined by careful selection of tubing internal diameter. The final mixture is then transferred to a specially designed debubbler which is connected to the cavity by a stainless-steel tube. The debubbler, which acts as a gas–liquid separator, ensures a continuous purging of gases into the cavity. The purging rate of the gas may be increased by introducing nitrogen from the bottom of the debubbler. The design of the debubbler prevents significant traces of solution or water vapours from entering the duct.

Table 3.3. Experimental Parameters for Continuous-flow Determination of Thiamine (Other Parameters as in Fig 3.15)

Parameter	Description
Cavity position	Flame centre, 26 mm above burner head
Photomultiplier voltage	900 V
Wavelength	384 nm
Slit width	2 mm (4 nm spectral bandpass)
Flow rates ($L\ min^{-1}$)	
Cooling water	0.1
Hydrogen	0.85
Nitrogen	1.60
Nitrogen carrier	0.010
Concentration, mol L^{-1}	
Sodium hydroxide	3.0
Orthophosphoric acid	3.0
Temperature, °C	
Delay coil	90
Mixing coil[1]	45
Time, s	
Sample	60
Wash	60

[1] Mixing coil (15 turns) between delay coil and debubbler. Reproduced by permission of the Royal Society of Chemistry from Ref. 39.

Table 3.3 provides information on the values of experimental parameters obtained in the determination of thiamine (42) using the analyser described above. The analysis proceeds automatically and is easier to perform compared to the corresponding manual determination. The successful determination of

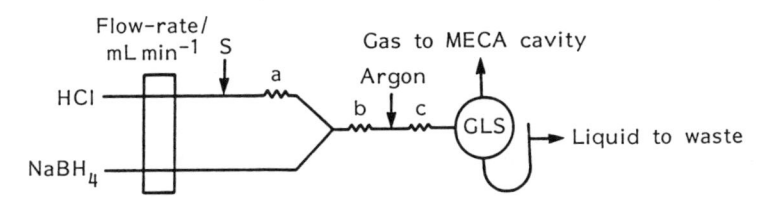

Figure 3.16. Flow injection manifold for the determination of arsenic with hydride generation and MECA detection. Coils a, b and c, 0.5 mm i.d.; S, point of sample injection; GLS, gas–liquid separator. Reproduced by permission of the Royal Society of Chemistry from Ref. 44.

Table 3.4. Experimental Parameters Used in FI–Hydride Generation–MECA System for the Determination of Arsenic

Experimental Parameter	Values
Sample volume (μL)	120
Coil b length (cm)	20
Coil c length (cm)	40
Solutions pumping rate (ml min^{-1})	1.5
Carrier gas flow rate (ml min^{-1})	0.5
Concentration, mol L^{-1}	
HCl in carrier solution	1.5
NaBH$_4$	0.2
EDTA in carrier solution	0.01
NaI in carrier solution	0.2
MECA parameters	
Wavelength (nm)	400
Slit width (nm)	1
Cavity	At flame centre
Cavity angle below horizontal	2°
Flow rates ($L\ min^{-1}$)	
Cooling water	0.01
Hydrogen	2.5
Nitrogen	4.5
Oxygen to cavity	0.90

Reproduced by permission of the Royal Society of Chemistry from Ref. 44.

many other sulphur-containing compounds in a variety of samples has also been accomplished (39,41,43). This procedure gives precise and accurate results, together with a satisfactory rate of sampling.

A flow injection molecular emission cavity analyser (FI–MECA) was designed by Burguera and Burguera for the determination of arsenic after generation of arsine (44). The configuration of this system is illustrated in Figure 3.16. The flow injection component of the system consisted of a five-channel peristaltic pump (Sage Orion Research) and a rotary injection valve (Rheodyne Model 7125). All tubing was made of Tygon (0.5 mm i.d.). The confluence T-joints and the connections were made of Perspex. A suitable gas–liquid separator, consisting of a narrow stainless-steel tube (i.d. 0.5 mm) was used for connecting the cavity to the flow injection system.

In this investigation, a water-cooled cavity was used as the detector. Arsine, generated in the FI system, was carried to the cavity by argon. During analysis (experimental conditions are shown in Table 3.4), the sample was injected via a rotary valve into a flowing stream of: 0.01 mol L^{-1} EDTA; 0.2 mol L^{-1} NaI;

and 1.5 mol L^{-1} HCl, which was then merged with one containing 0.2 mol L^{-1} NaBH$_4$. After gas–liquid separation, the arsine was purged into the cavity and the emission was recorded at 400 nm. This method was rapid, simple and very sensitive for determining arsine (see Chapter 4).

3.5 CONCLUSIONS

The instruments described in this chapter are not now commercially available. As a consequence, the technique is not widely used and new researchers are not aware of how easily a MECA photometer can be constructed. However, it should be clear from the descriptions of the various MECA instruments referred to in this chapter that widely available units commonly used in photometry can be used to assemble a simple MECA instrument, and that this can be carried out even by laboratories with limited research budgets.

REFERENCES

1. A. C. Calokerinos and T. P. Hadjiioannou, *Anal. Chim. Acta*, **148**, 277 (1983).
2. S. V. Wanegen and Q. Fernando, *Anal. Chem.*, **57**, 2743 (1985).
3. N. Grekas, PhD Thesis, University of Athens, 1988.
4. S. L. Bogdanski, M. Burguera, and A. Townshend, *CRC Crit. Rev. Anal. Chem.*, **10**, 185 (1981).
5. A. C. Calokerinos, PhD Thesis, University of Birmingham, 1978.
6. R. Belcher, S. L. Bogdanski, and A. Townshend, *Anal. Chim. Acta*, **67**, 1 (1973).
7. R. Belcher, S. L. Bogdanski, R. A. Sheikh, and A. Townshend, *Analyst*, **101**, 562 (1976).
8. Th. A. Kouimtzis, *Anal. Chim. Acta*, **88**, 303 (1977).
9. R. Belcher, S. L. Bogdanski, D. J. Knowles, and A. Townshend, *Anal. Chim. Acta*, **77**, 53 (1975).
10. O. Osibanjo, PhD Thesis, University of Birmingham, 1976.
11. R. Belcher, S. L. Bogdanski, Z. M. Kassir, and A. Townshend, *Anal. Lett.*, **7**, 751 (1974).
12. R. Belcher, S. L. Bogdanski, S. A. Ghonaim, and A. Townshend, *Nature*, **248**, 326 (1974).
13. C. O. Akpofure, PhD Thesis, University of Birmingham, 1976.
14. R. Belcher, S. L. Bogdanski, A. C. Calokerinos, and A. Townshend, *Analyst*, **106**, 625 (1981).
15. R. Belcher, S. A. Ghonaim, and A. Townshend, *Anal. Chim. Acta*, **71**, 255 (1974).
16. I. Z. Al-Zamil, PhD Thesis, University of Birmingham, 1978.
17. S. Liawruangrath, PhD Thesis, University of Birmingham, 1980.
18. S. L. Bogdanski, E. Henden, and A. Townshend, *Anal. Chim. Acta*, **116**, 93 (1980).
19. M. Burguera, S. L. Bogdanski, and A. Townshend, *Anal. Chim. Acta*, **153**, 41 (1983).
20. R. Belcher, S. L. Bogdanski, E. Henden, and A. Townshend, *Anal. Chim. Acta*, **113**, 13 (1980).
21. R. Belcher, S. L. Bogdanski, A. C. Calokerinos, and A. Townshend, *Analyst*, **102**, 220 (1977).

22. A. Celik and E. Henden, *Analyst*, **114,** 563 (1989).
23. J. D. Flanagan and R. A. Downie, *Anal. Chem.*, **48,** 2047 (1976).
24. S. L. Bogdanski, A. C. Calokerinos, and A. Townshend, *Int. Lab.*, **12,** 66 (1982).
25. M. Burguera, J. L. Burguera, and M. Callignani, *Anal. Chim. Acta*, **138,** 137 (1982).
26. M. Burguera and J. L. Burguera, *Anal. Chim. Acta*, **153,** 53 (1983).
27. R. Belcher, S. L. Bogdanski, E. Henden, and A. Townshend, *Anal. Chim. Acta*, **92,** 33 (1977).
28. E. Henden, *Anal. Chim. Acta*, **173,** 89 (1985).
29. M. J. Cope and A. Townshend, *Anal. Chim. Acta*, **134,** 93 (1982).
30. I. H. El-Hag, *Anal. Proc.*, **19,** 320 (1982).
31. J. K. Foreman and P. B. Stockwell, (Eds), *Topics in Automatic Chemical Analysis*, Ellis Horwood Publishers, Chichester, 1979.
32. J. Bierens de Haan, in J. K. Foreman and P. B. Stockwell (Eds), *Topics in Automatic Chemical Analysis*, Ellis Horwood Publishers, Chichester, 1979.
33. I. H. El-Hag and A. Townshend, *J. Anal. Atom. Spectrosc.*, **1,** 383 (1986).
34. N. P. Evmiridis and A. Townshend, *J. Anal. Atom. Spectrosc.*, **2,** 339 (1987).
35. J. L. Burguera and M. Burguera, *Anal. Chim. Acta*, **157,** 177 (1984).
36. J. L. Burguera and M. Burguera, *Anal. Chim. Acta*, **170,** 331 (1985).
37. Burguera J. L., and Burguera M., *Anal. Chim. Acta*, **179,** 497 (1985).
38. J. Ruzicka, *Anal. Chim. Acta*, **261,** 3 (1992).
39. N. Grekas and A. C. Calokerinos, *Analyst*, **110,** 335 (1985).
40. N. Grekas and A. C. Calokerinos, *Anal. Chim. Acta*, **173,** 311 (1985).
41. N. Grekas and A. C. Calokerinos, *Anal. Chim. Acta*, **225,** 359 (1989).
42. N. Grekas, A. C. Calokerinos, and T. P. Hadjiioannou, *Analyst*, **114,** 1283 (1989).
43. N. Grekas and A. C. Calokerinos, *Analyst*, **115,** 613 (1990).
44. M. Burguera and J. L. Burguera, *Analyst*, **111,** 171 (1986).

CHAPTER 4

SULPHUR, SELENIUM AND TELLURIUM

E. HENDEN

Ege University
Izmir, Turkey

4.1.	**Introduction**		71
4.2	**Sulphur Compounds**		72
	4.2.1	Determination of Sulphur by Conventional MECA	73
		4.2.1.1 Inorganic Sulphur Compounds	73
		4.2.1.2 Organic Sulphur Compounds	76
		4.2.1.3 Sulphur Compounds in Detergents	79
		4.2.1.4 Sulphur in Solids	80
		4.2.1.5 Indirect Determinations Based on the S_2 Emission	83
		4.2.1.6 Automated Conventional MECA	85
	4.2.2	Gas Generation Systems	85
		4.2.2.1 Determination of Sulphur	85
		4.2.2.2 Automated Gas Generation Systems	87
	4.2.3	Determination of Sulphur Compounds after Gas and Liquid Chromatographic Separation	89
4.3	**Selenium and Tellurium Compounds**		91
	4.3.1	Determination of Selenium and Tellurium by Conventional MECA	92
		4.3.1.1 Inorganic Selenium and Tellurium Compounds	92
		4.3.1.2 Organic Selenium and Tellurium Compounds	94
	4.3.2	Determination of Selenium and Tellurium by Gas Generation Systems	94
4.4	**Conclusions**		96
References			96

4.1 INTRODUCTION

Organic and inorganic sulphur compounds which are widely used in industry also exist in nature. It is, therefore, important to be able to analyse for them at trace and subtrace levels. The sensitivity of MECA for sulphur compounds in gaseous, liquid and solid samples is based on the principle of chemiluminescence, so it is ideally suited for their determination.

Flame Chemiluminescence Analysis by Molecular Emission Cavity Detection
Edited by D. A. Stiles, A. C. Calokerinos and A. Townshend   © 1994 John Wiley & Sons Ltd

Conventional MECA has been used for the speciation of sulphur, selenium and tellurium compounds in a variety of matrices where various approaches have been described for the reduction of matrix effects and interferences. However, the full potential of MECA is realized in gas generation systems where chemiluminescence becomes more effective. Further modifications have been described in the use of the cavity as a detector for gas and liquid chromatography.

4.2 SULPHUR COMPOUNDS

The first paper describing the MECA phenomenon by Belcher et al. (1) described the determination of compounds sulphur. Since then, these compounds have been extensively studied by MECA. All sulphur-containing compounds give a characteristic blue emission of S_2 molecules in the MECA cavity when a hydrogen diffusion flame is used. The most intense peaks of the multi-peaked band emission occur between 350 and 410 nm (see Chapter 2), and the emission intensity is usually measured at 384 nm (1,2).

This emission intensity varies with flame composition and temperature. However, the addition of air to the H_2/N_2 flame gives rise to an unusual phenomenon. At first, the sulphur emission is destroyed, but on increasing the air pressure the emission from the cavity reappears and becomes even more intense than in the absence of air (1,3). The initial reduction of intensity has been attributed to an increase in flame temperature which destroys the S_2 molecules to form sulphur–oxygen compounds (1). However, as the amount of oxygen added to the flame increases, it reacts stoichiometrically with hydrogen at the burner top and the S_2 emission appears bright again (4).

The emission intensity, (I), of the S_2 species generated within a MECA cavity is related to the concentration of sulphur, (C), by the equation:

$$I = KC^n$$

where K and n are constants which depend on the experimental conditions and nature of the sulphur compound being investigated. (5,6) (see Chapter 2). The calibration curve becomes linear on plotting log I vs. log C (7).[1] The upper curvature of the graph has been attributed to self-absorption or to the formation of polyatomic sulphur species (S_4, S_8, etc.) (2).

Another explanation of the negative curvature of the graph at high sulphur concentrations is a change in the chemiluminescent reaction mechanism (1). This observation is verified by the appearance of multipeaked responses at high concentrations while one peak is observed at low sulphur concen-

---

[1] For simplicity throughout this chapter the log I vs log C curve will be called a log–log calibration graph, while the exponential curve will be referred to as the calibration graph.

trations. The effect arises from interaction of some of the initial decomposition products with undecomposed sample to form a less volatile product, together with deposition of sulphur (6). Analytical reliability is best when the amount of sample in the cavity is small enough to produce only a single-peaked response for a given compound.

4.2.1 Determination of Sulphur by Conventional MECA

4.2.1.1 Inorganic Sulphur Compounds

In conventional MECA, using a stainless-steel cavity, the t_m value (See Chapter 2) increases with the thermal stability of the sulphur anion (Table 4.1). The t_m values are also influenced by the presence of cations in the sample. Alkali metals, for example, form particularly stable sulphates, greatly delay the appearance of the S_2 emission, and depress the peak height (2). Sulphates with high t_m values generate weak emissions because, at the time of their appearance, the cavity surface is hot and the effect of the Salet phenomenon (see Chapter 2) is minimal. The peaks are broad due to the low rate of temperature increase of the cavity and the relatively low volatility of the analyte. Thus, as shown in Figure 4.1, the t_m value of sulphuric acid is lower than that of the

Table 4.1. MECA Characteristics of Some Inorganic Sulphur Compounds in a Stainless-steel Cavity Introduced into an H_2/N_2 Flame. Gas Flow Rates: H_2, 1.7 L min^{-1}; N_2, 4.0 L min^{-1}

Compound	t_m (s)	Detection Limit (ng sulphur/5 µL)	Slope of Log–Log Plot
With silica liner			
Na_2S	2.0	5[1]	1.4
KSCN	18.0	5[1]	2.0
Na_2SO_3	2.4	6[2]	1.9
$Na_2S_2O_4$	2.0, 10.0	10[3]	1.7
Na_2SO_4	40.0	20[2]	1.8
$Na_2S_2O_8$	40.0	20[2]	1.6
Without silica liner			
S	1.5	0.4	1.2
$Na_2S_2O_4$	7.5	5[1]	1.1
$Na_2S_2O_6$	11, 20	1.5, 5[2]	2.1, 0.8[2]

[1] 0.006 mol L^{-1} $(NH_4)_2HPO_4$.
[2] 0.1 mol L^{-1} H_3PO_4.
[3] 0.020 mol L^{-1} phosphate buffer.
Reprinted with permission from Burguera *et al.*, *CRC Crit. Rev. Anal. Chem.*, **10**, 185 (1980). Copyright CRC Press, Inc., Boca Raton, FL, USA.

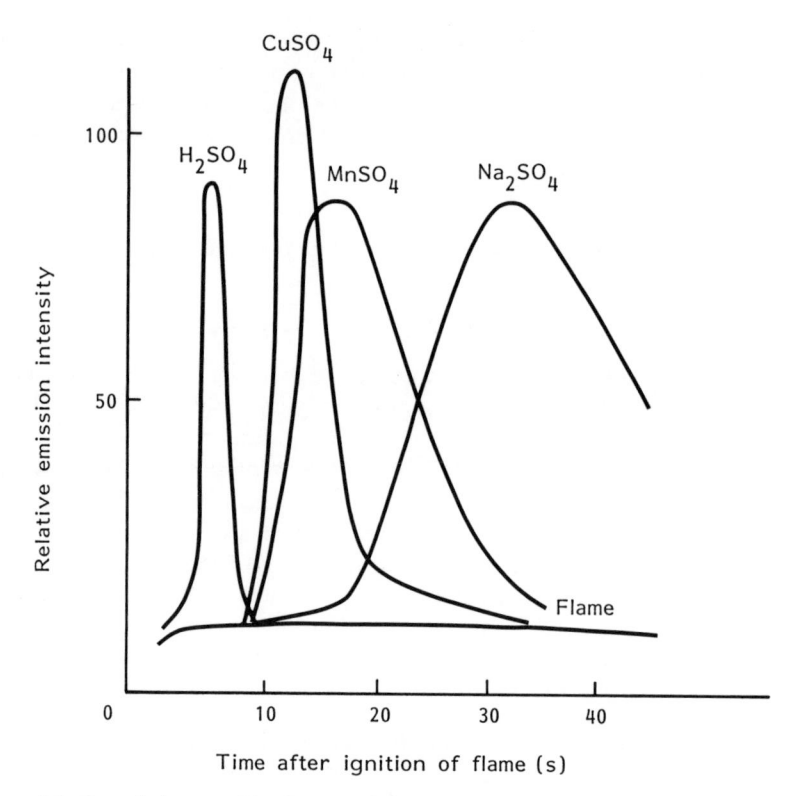

Figure 4.1. S_2 emission profiles from various sulphates. Reproduced by permission of Elsevier Science Publishers, BV, from Ref. 1.

metal sulphates. Responses other than from pure sulphuric acid are not analytically useful due to reduced sensitivity and overlapping of peaks.

The interference of metal ions on the emission from sulphates is reduced or eliminated by the addition of phosphoric acid to the sample solution (8–10). The response obtained is identical to that from a binary mixture of sulphuric and phosphoric acids. The concentration of phosphoric acid influences the extent to which the cation interference is reduced, provided that the concentration of acid is higher than 0.01 mol L^{-1} (9,10). Thus, the most intense S_2 emission intensity from 0.01 mol L^{-1} iron(II) sulphate is obtained by the addition of 0.07 mol L^{-1} phosphoric acid while higher concentrations of phosphoric acid leave the sulphate response unaffected (8). The same results were obtained for manganese, sodium and ammonium sulphates (8).

The slope of log–log calibration graphs for sulphuric acid decreases with increasing phosphoric acid concentration. Thus, the log–log calibration graph has slopes of 1.7 and 1.3 for the MECA determination of sulphur as sulphuric acid in the presence of 0.04 and 0.4 mol L^{-1} phosphoric acid, respectively

(9). As the t_m value increases, the slope of the calibration graph decreases. Bogdanski et al. (9) have attributed this effect to the delayed heating of the sample, because of its dispersion in phosphoric acid.

Cardwell et al., using a silica cavity (10), reported that residual phosphate species in the cavity enhance the S_2 emission but at the same time decrease its lifetime. Hence, in order to avoid problems associated with having residual phosphate in the cavity, scrupulous cleaning before re-use was recommended (9).

Other sulphur anions are affected by the presence of cations but to a much lesser extent than sulphate. Phosphoric acid at a concentration of 0.01 mol L^{-1} has been recommended for removing cationic interferences in the determination of peroxodisulphate, sulphite and thiocyanate ions (8). However, it cannot be used for the determination of sulphide and thiosulphate because hydrogen sulphide is formed during sulphide analysis and is lost, and thiosulphate disproportionates. However, diammonium hydrogen phosphate buffer solution at pH 7.0 is satisfactory and has been recommended for use in the analysis of these two anions (8). Using the procedures outlined, sulphur anions have been determined with high sensitivity (8,11,12) (Table 4.1).

This technique has also been used successfully in determining milligram per litre levels of sulphate in high purity water (13,14) and in aqueous extracts of soils, dusts, vinegar and charcoal (14), without tedious sample pretreatment. The results compare well with those obtained by nephelometry but the MECA procedure is faster and more sensitive.

The determination of inorganic sulphate in urine has been reported (15). Urine is diluted 100-fold with 0.1 mol L^{-1} phosphoric acid, injected into a carbon cavity, and introduced into a hydrogen–nitrogen–air flame. The sharp sulphate peak allows direct determination of sulphate with 3–5% repeatability.

The differences in t_m values (Table 4.1) means that it is possible to display resolved peaks from nearly all binary and a number of ternary mixtures (Figure 2.16) of sulphur anions by selecting appropriate flame conditions (11,12). This provides a simple means for speciating common sulphur anions.

Some sulphur anions give multi-peaked emissions which are attributed to their mode of breakdown. One example is thiosulphate which, in the presence of phosphate buffer and when using a hydrogen–nitrogen flame, gives two peaks with t_m values equal to 2 and 10 s. The breakdown mechanism (12) which takes place is:

$$Na_2S_2O_3(l) \longrightarrow Na_2SO_3(s) + S(s)$$

Thus, in the order of increasing t_m value, the two peaks correspond to sulphur and sulphite. If air is also introduced into the flame, a three-peak sequence appears with t_m values equal to 2, 4 and 8 s. These three peaks apparently arise from the decomposition of sodium thiosulphate (12), which follows the reaction sequence:

$$4Na_2S_2O_3 \cdot 5H_2O \longrightarrow 4Na_2S_2O_3(l) + 20H_2O(g)$$
$$4Na_2S_2O_3(l) \longrightarrow 3Na_2SO_4(s) + Na_2S_5(s)$$
$$Na_2S_5(s) \longrightarrow Na_2S(s) + 4S(l)$$

In the order of increasing t_m value, the three peaks would correspond to sulphide, sulphur and sulphate.

Thiocyanate gives a non-symmetrical peak in the presence of orthophosphoric acid. The observation has been attributed to the depolymerization of thiocyanic acid which is formed on the addition of this acid. Dithionite gives at least four peaks, three of which can be assigned to sulphur or sulphide, sulphite and sulphate, in agreement with the suggested decomposition mechanism of dithionite where thiosulphate, sulphite and sulphur dioxide are formed (2):

$$2S_2O_4^{2-} \longrightarrow S_2O_3^{2-} + SO_3^{2-} + SO_2(g)$$

Thiosulphate is then decomposed as described above. Dithionate generates two peaks, the first ascribed to sulphur dioxide and the second to sulphate, following the breakdown mechanism below (2):

$$S_2O_6^{2-} \longrightarrow SO_2(g) + SO_4^{2-}$$

Wagenen and Fernando (16) employed a spinning sample cup and adjustable jet flame (see Chapter 3) for the determination of sulphur anions. The analyte solution was injected into a silica cavity at the end of a rotating metal rod and then the emission was recorded. Sulphite and sulphate were resolved very well but the reproducibility for sulphate was much better than that for sulphite. This was attributed to the loss of the easily decomposed sulphite during the introduction of the cavity into the flame.

Resolution of the MECA responses from various sulphur anions has been used to determine sulphite and sulphate in wine (17). A scheme has also been proposed for the determination of sulphide, sulphate, thiocyanate and thiosulphate ions in effluents (17,18). Ions are removed by treatment of the effluent with selective precipitating agents. Those that remain in solution are oxidized to sulphate with hydrogen peroxide. The sulphate is then quantified by MECA. In this way the concentration of each ion present in the original effluent sample can be determined either directly (SCN^-), or by difference (S^{2-}, SO_4^{2-}, $S_2O_3^{2-}$).

4.2.1.2 Organic Sulphur Compounds

The S_2 emission is suitable for determining organosulphur compounds at the sub-nanogram level by conventional MECA. The emission intensity depends

Table 4.2. Analytical Characteristics of Various Organic Sulphur-Containing Compounds by Conventional MECA (20)

Compound	t_m s	Linear Range ng S	Limit of Detection ng S/5 μL
Sulphadiazine	2.0	15–120	2.0
Sulphamerazine	2.1	15–150	2.0
Promethazine	0.2	0.4–4.0	0.2
Perphenazine	0.2	0.5–4.0	0.3
Sulphathiazole	0.3, 1.2, 2.1	2–12	0.7
Acetazolamide	0.9, 1.2	10–50	1.5
Cystine	0.5	2–17	1.2
Cysteine	0.5	2–15	1.0
Taurine	1.4	10–50	4.0
Methionine	0.25	1–12	0.6
Penicillamine	0.25	1–10	0.6
Glutathione	0.5	2–20	1.2

on the nature of the compound, the sensitivity decreasing as the t_m value increases, with the limit of detection being very low for volatile compounds. The emission intensity is reduced by organic solvents unless they are removed before the S_2 emission is generated. This effect has been attributed to the consumption of hydrogen atoms by organic fragments, which are not present in a purely aqueous solvent systems (1,19).

Sulphadiazine, sulphamerazine and sulphathiazole generate a single emission with a t_m of approximately 2 s (20). The relatively low MECA sensitivity obtained for these sulphonamides has been attributed to the strength of the S=O bond (498 kJ mol-1) (2) which is less easily ruptured than the weaker S–C bond (272 kJ mol-1), especially with the cool flame conditions used.

Promethazine, perpherazine and acetazolamide generate an S_2 emission with very short t_m values (0.2 s). Thus, when emission occurs, the cavity is relatively cool, the Salet phenomenon comes into effect, and very sensitive determinations of these triazines may be made (Table 4.2.) (2,20).

Sulphonamides and triazines offer an excellent example of how the cavity material affects emission intensity. Triazines are derivatives of thiodiphenylamine, which decomposes at 371°C. Nevertheless, the corresponding S_2 emission appears when the temperature of the cavity is just above 100°C. Thus, it is assumed that the early appearance of this emission is due to a catalytic effect of the steel surface of the cavity on the chemiluminescent reaction of the compound with hydrogen atoms from the flame (2). This suggestion is verified by the different behaviour of the same compounds in a carbon cavity under the same experimental conditions. For example, promethazine gives a

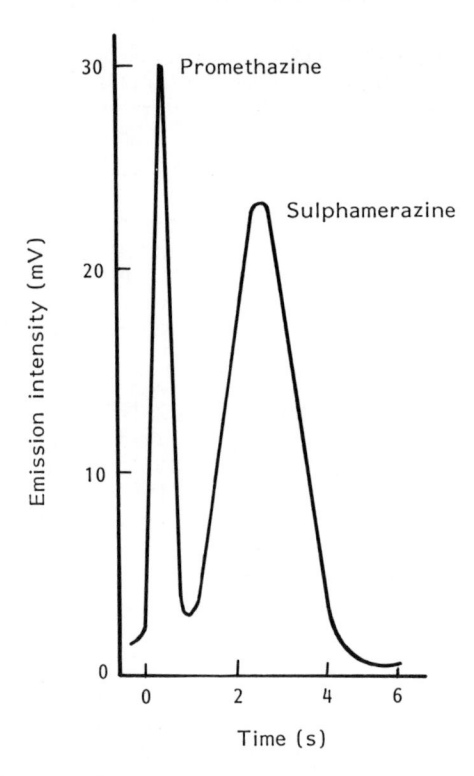

Figure 4.2. Resolution of a mixture of promethazine hydrochloride (3 ng S) and sulphamerazine (40 ng S) using a stainless-steel cavity after evaporation of acetone. Reproduced by permission of Elsevier Science Publishers, BV, from Ref. 12.

fast emission (t_m = 0.2 s) in a stainless-steel cavity, whereas in a carbon cavity, two peaks of t_m = 0.2 s and 1.5 s are obtained and the sensitivity is reduced by at least 33 times (2).

A mixture of promethazine and sulphamerazine gives resolved MECA peaks in a stainless-steel cavity (Figure 4.2). The amount of triazine does not affect the peak height of the sulphonamide, but sulphamerazine containing more than 40 ng of sulphur suppresses the emission from promethazine, presumably by interfering with the decomposition mechanism (2).

Common matrix ingredients of tablets like starch, lactose and magnesium stearate are in such high concentration compared to the drug itself that they suppress the S_2 emission and must therefore be removed. This is done by grinding up the tablet and extracting the active ingredient with either acetone or, in the case of heterocyclic drugs, n-hexane after dissolution in water at pH 11. Measurement of the S_2 emission intensity then allows determination of the drug's concentration directly.

Saccharin in the range 20–500 μg mL^{-1}, with a limit of detection of 2 μg mL^{-1}, has been determined successfully in soft drinks (21). The interfering components of the sample were removed from alkaline solution by extraction into diethyl ether. The aqueous solution was then acidified and the saccharin extracted into ethyl acetate. Aliquots of the organic phase were then injected into the cavity for generation and measurement of the S$_2$ emission.

Other organic sulphur compounds which have been determined by MECA include vitamins (triamine, B1 and biotin, H) (22), amino acids (methionine, cysteine, cystine, taurine, penicillamine and glutathione) and proteins (Table 4.2) (20).

Inorganic sulphate and sulphur-containing proteins in egg albumin and serum albumin preparations give resolved MECA peaks with t_m = 0.2 s and 0.8 s for protein, and 2.4 s for sulphate. These two compounds can, therefore, be differentiated easily by MECA (22).

4.2.1.3 Sulphur Compounds in Detergents

For determining sulphates and sulphonates in detergents, carbon cavities rather than those made of other materials have been used (23). Organic sulphonates (RSO$_3{}^-$) give multi-peaked responses. Sodium dodecylbenzene sulphonate generates two peaks: the first is sharp with t_m = 2 s and the second broad with t_m = 36 s, and similar to that given by sodium sulphate. When 0.1 mol L^{-1} phosphoric acid is added to the analyte solution, three unresolved peaks are obtained, the last one having the same t_m value as that of sodium sulphate in 0.1 mol L^{-1} phosphoric acid. By using the second most intense peak, 2.5 ng of sulphur can be detected. The behaviour of sodium hexadecane sulphonate and p-xylene-2-sulphonic acid in 0.1 mol L^{-1} phosphoric acid are similar.

Organic sulphates (ROSO$_3{}^-$) such as sodium lauryl sulphate give only a weak, delayed response, similar to that from inorganic sulphates. In the presence of 0.1 mol L^{-1} phosphoric acid, these same organic sulphates generate more intense peaks with lower t_m values, similar to those observed with sodium sulphate, as well as a third peak of sodium dodecylbenzene sulphonate. The similar behaviour of organic and inorganic sulphates with phosphoric acid is attributed to hydrolysis of organic sulphates by this acid in the cavity (2).

Commercial detergents may contain inorganic and organic sulphates and sulphonates. An analysis scheme for the determination of these substances in detergents by MECA has been described (17,23). In this scheme the sample is combusted in an oxygen flask containing aqueous hydrogen peroxide. After acidification with phosphoric acid and appropriate dilution, total sulphur is measured by injecting 5 μL aliquots into the cavity. For the determination of total organic and inorganic sulphates, the detergent is boiled for 25 min with concentrated phosphoric acid to hydrolyse organic sulphates to sulphuric acid. After extracting the sulphonates in the solution into carbon tetrachloride with p-toluidine, sulphate is determined in the aqueous phase. The original inorganic sulphate is determined in the aqueous phase after extracting unhydrolysed organic sulphates and sulphonates with p-toluidine. The sulphonate present is calculated by difference.

The p-toluidine ion-pair extraction procedure is not sensitive enough for the analysis of sulphonates in river water, etc. However, ion-pair extraction with methylene blue followed by solvent evaporation in the cavity, can be used to concentrate sulphonates before the final measurement (2). The procedure has a limit of detection equal to 10 ng of sulphur.

4.2.1.4 Sulphur in Solids

Analysis of sulphur species in solids is of great importance since the analytical procedure does not involve a solubilization step and is, therefore, less time consuming. Moreover, when the concentration of each sulphur species rather than total concentration of the sulphur is required, as is usually the case in environmental and toxicological analysis, unavoidable conversion of one type of species to another through oxidation or reduction processes may take place during solubilization.

Direct analysis of solid samples by MECA has been investigated by Calokerinos and Townshend (17). In their study, a few milligrams of the solid sample were placed in the cavity and the emission was recorded conventionally. The approach appeared to be promising and they identified several criteria for a successful analysis:

1. the matrix composition of standards and samples must be matched.
2. the amount of sample placed in the cavity should be small in order to avoid severe built up of deposits.
3. the samples must be finely-ground to obtain a homogeneous analyte powder.
4. A wetting agent must be added to the sample after transferring it to the cavity so that the powder is spread out homogeneously on its inner surface.

Solid sample introduction into the cavity has been applied to the determination of sulphur drugs in tablets (22). Several micrograms of sample, ground and sieved through 60-mesh sieve, were added to a carbon cavity and the sample wetted with acetone. After 45 s, when the acetone had evaporated, the cavity was rotated into a hydrogen–nitrogen–air flame. As soon as the emission had ceased, the temperature of the cavity was increased by turning off the nitrogen supply, any residue was allowed to volatilize, the cavity was cooled to room temperature, and then used for the next measurement. The t_m values were the same as when solutions were injected, but the emission intensities were less than from pure drugs, mainly because co-existing ingredients, such as starch and magnesium stearate in the tablets, had a depressive effect.

Weighing very low amounts of sample is a difficult task and tends to lead to poor accuracy and precision. Therefore, as an alternative to the above procedure, an aluminium block with a groove at one end, capable of accommodating a known weight, e.g. 250 µg, of sample, was used to transfer the sample into the cavity (24). Thus, sulphur was determined in iron ores and blast furnace slags with a reproducibility of 6.5%. The same procedure was also applied to sulphur in pitch and lichen samples (2).

The sulphur content of coal has also been studied by MECA (25). Each type of sulphur-bearing constituent, namely organic, pyritic and sulphate sulphur gave a resolved peak. Powdered coal, suspended in acetone, was placed in the cavity, the acetone allowed to evaporate, and the cavity heated. The first emission was ascribed to pyritic sulphur and various organic sulphur compounds, the second to pyritic sulphur alone, and the last to inorganic (mainly calcium) sulphate (Figure 4.3). Rapid determination of these forms of sulphur in coal is, therefore, possible when the suitable standard coal samples are used (25).

Sulphur in teeth has been determined by pressing 1 to 3 mg of powdered dentine into an iron or titanium cavity which was then introduced into the flame (26). Two groups of peaks were obtained, one due to sulphur(II) (cysteine, cystine and methionine) and the other due to sulphur(VI) (chondroitin sulphate). Sulphur in the range 0.01–0.03% could be determined with a limit of detection equal to 0.004%.

Fernando et al. (27,28) has described the identification and simultaneous determination of sulphite and sulphate in solids. A solid sample weighing between 0.2 and 3.0 mg was placed in a small aluminium cup and then wetted with 10 µL of a solution containing 0.10 mol L^{-1} phosphoric acid and 0.005% (v/v) Triton X-100. The cup was placed in a quartz-lined cavity, heated, and the emission intensity at 384 nm measured. After completion of each measurement, the aluminium cup was discarded. Sodium sulphite and sodium sulphate were each dispersed in silicon dioxide to give mixtures containing about 500 mg sulphur per kilogram. Standards were then prepared by mixing appropriate amounts of these suspensions. The calibration graph was linear over the range 0.1–10 µg of sulphur for either sulphite or sulphate. Silicon dioxide was chosen as the major matrix component since it does not interfere with the measurement.

Sulphite, elemental sulphur, sulphide and sulphate were detected and quantified in solid samples with limits of detection ranging from 0.2 pg for sulphite to 8 ng for sulphate in a 1.5 mg sample (28). In the case of sulphate, however, an argon-cooled hydrogen flame was reported to give better sensitivity than a hydrogen flame cooled with nitrogen. Reasons given for this behaviour were its higher temperature and a decreased quenching of the excited S_2 species (29). It was suggested that the argon-cooled flame could be used in the determination of mixtures containing inorganic sulphur species in solids within the

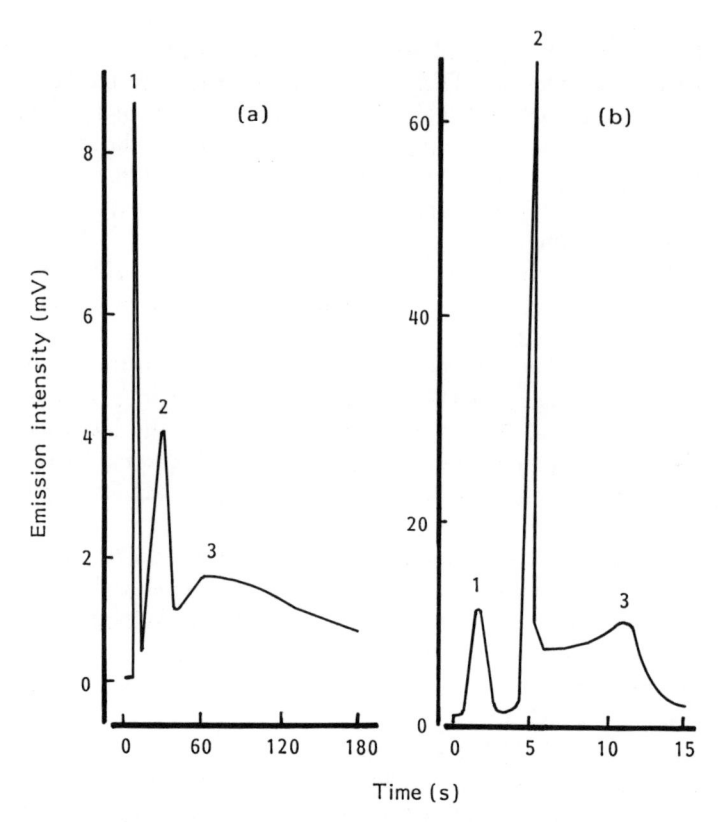

Figure 4.3. S_2 emission profiles from (a) a mixture of solids (1) thiourea, (2) calcium sulphide, (3) calcium sulphate, and (b) powdered coals showing responses from (1) organic and pyritic sulphur, (2) pyritic sulphur, and (3) sulphate sulphur. Reprinted with permission from Burguera *et al.*, *CRC Crit. Rev. Anal. Chem.*, **10**, 185 (1980). Copyright CRC Press, Inc., Boca Raton, FL, USA.

range 10 to 5000 mg kg^{-1}. However, it should be noted that the reproducibility was poor and varied between 10 and 20%. Poor reproducibility was attributed to inhomogeneity of the small solid standards (1–2 mg), variable surface composition of the aluminium sample cup, and trace sulphate impurities present in the silicon dioxide which was used to prepare standards.

Kouimtzis (30) has determined sulphur dioxide in air by using a modified stainless-steel cavity. The sulphur dioxide was absorbed by a small amount (20–40 mg) of silica gel which was then deposited into a suitably designed cavity. This was then heated in a hydrogen–nitrogen–air flame to obtain the S_2 emission. The calibration curve of emission intensity against concentration covered the range 5–120 ng of sulphur dioxide. Sulphuric acid mist and sulphur-containing organic compounds also interfered by giving S_2 emissions;

water vapour and organic vapours interfered by quenching the S_2 emission. These compounds were, therefore, removed from the absorption train before the sulphur dioxide was absorbed into silica gel. Sulphuric acid mist was removed by a glass fibre filter paper, mercaptans by copper sulphate on pumice, water vapour through absorption into magnesium perchlorate, and organic compounds were trapped by paraffin wax. The results obtained by this procedure compared well with those of the well-known West–Gaeke method, but the MECA procedure required a much smaller air sample (0.5 L compared to 20 L).

Determination of sulphur in high purity selenium has been achieved by using a procedure similar to that of Kouimtzis for the determination of sulphur dioxide in air (31). The sample containing 2.5 to 60 ng of sulphur was first combusted at 800°C in a stream of oxygen. The sulphur dioxide evolved was absorbed by silica gel which was then introduced into the flame to determine the S_2 emission response. Selenium dioxide interfered by generating an Se_2 emission. However, this was removed by absorption in a trap containing glass wool and glass fibre filter paper.

A filter cavity has been designed (32) in which solids like precipitates or air particulates can be collected within the cavity before it is used for direct emission measurements. This cavity eliminates problems associated with the transfer of filters from the collection device to the cavity.

4.2.1.5 *Indirect Determinations Based on the S_2 Emission*

Various compounds that do not contain sulphur have been determined by first converting them to sulphur-containing derivatives and then measuring the S_2 emission of the derivatives. The following examples indicate where this technique has been used most successfully, but more definitive information may be found under the appropriate headings in subsequent chapters.

Carbonyl compounds have been determined from S_2 emissions after forming sulphite addition compounds with excess sodium sulphite (33):

$$\underset{R'}{\overset{R}{}}\!\!\!\!\!C=O + SO_3^{2-} + H_2O \longrightarrow \underset{R'}{\overset{R}{}}\!\!\!\!\!C\underset{SO_3^-}{\overset{OH}{}} + OH^-$$

$$\xrightarrow{+ H^+} \underset{R'}{\overset{R}{}}\!\!\!\!\!C\underset{SO_3H}{\overset{OH}{}}$$

Alcohols such as vic-diols, α-aminoalcohols (33) and other polyhydroxy compounds (18) may be first converted to carbonyl compounds by periodate oxidation (34):

$$CH_2OH(CHOH)_4CH_2OH + 5IO_4^- \longrightarrow 2HCHO + 4HCOOH$$
$$+ 5IO_3^- + H_2O$$

and then determined from the S_2 emission of a sulphite addition compound. Ethanol can also be determined (35) by first oxidizing ethanol to acetaldehyde using nicotinamide adenine dinucleotide in the presence of yeast alcohol dehydrogenase at pH 8 and then converting the resulting acetaldehyde to its sulphite addition compound, which is then injected into the cavity.

Cyanide has also been determined indirectly through the S_2 emission (36). In this case the ability of this anion to release sulphite from the sulphite-formaldehyde addition compound is utilized:

$$CH_2(OH)SO_3^- + CN^- \longrightarrow CH_2(OH)CN + SO_3^{2-}$$

Safavi and Townshend (37) have described an indirect method for the determination of amines by their reaction with the formaldehyde–sulphite addition compound. The amine solution is added to a mixture of sulphite and formaldehyde that has been allowed to react for 15 min. After addition of phosphoric acid, the reaction product is measured by injecting 5 μL of the mixture into a carbon cavity and monitoring the S_2 emission. For aromatic amines, the mixture is set aside for one hour to allow the reaction to go to completion before the solution is injected into the cavity.

Aliphatic amines and amino acids can be converted to their diethyldithiocarbamates by reaction with carbon disulphide (38), while amino acids may also be determined after reaction with 2,4,6-trinitrobenzene-1-sulphonic acid, (TNBS), at pH 11.7 (39).

The low temperature of the hydrogen–nitrogen flame used in MECA is not sufficient to stimulate emission from metals when their salts are injected directly into the cavity. Nevertheless, a variety of metals have been determined indirectly by measuring the S_2 emission from chelates with sulphur-containing ligands (40).

Arsenic, antimony, selenium and tellurium can be measured directly by MECA but indirect methods have been described by Safavi and Townshend (41). The elements are first extracted into tetrachloromethane as their diethyldithiocarbamates and then 5 μL aliquots of the extract are injected into a stainless-steel cavity. The S_2 emission intensity from the extracted chelate is subsequently measured. Organic solvents suppress the S_2 emission but this effect is eliminated by using a cold air current to blow the solvent from the cavity prior to measurement. Interferences from other metals which can be extracted as dithiocarbamates are avoided by EDTA masking.

4.2.1.6 Automated Conventional MECA

The reproducibility of MECA is usually dependent on the operator's skill in controlling the experimental conditions. Therefore, reproducibility is improved by automation because the experimental variables are controlled better. This has been proved by El-Hag and Townshend (42) who described an automated system for sample injection and cavity rotation into the flame (see Chapter 3). Evmiridis and Townshend (43) have evaluated this system for the determination of thiourea, sulphuric acid, ammonium thiocyanate and promethazine.

Burguera and Burguera (44) have developed a flow injection system for the conventional MECA determination of sulphide, sulphite and sulphate at the nanogram level. A water-cooled stainless-steel cavity was held continuously in the flame. The sulphur-containing species were injected as 3 µL aliquots into the carrier stream, which was then directed into the cavity for generation and measurement of the S_2 emission intensity. The appearance time of the peaks (t_m values) increased in the order sulphide, sulphite and sulphate, which, therefore, allowed the simultaneous determination of the three sulphur species in admixture. By changing the carrier stream from water to hydrogen peroxide only one peak was obtained from the ternary mixture, that corresponding to sulphate. Thus, it was possible to determine total sulphur in a sulphur-containing mixture.

4.2.2 Gas Generation Systems

4.2.2.1 Determination of Sulphur

A gas generation system for the determination of sulphide has been developed (45). Hydrogen sulphide, generated by injecting the sulphide solution into a reaction vessel containing 2.0 mL of 3–8 mol L^{-1} hydrochloric acid, was swept by nitrogen directly into a water-cooled stainless-steel cavity continually held in a hydrogen–nitrogen flame. A drying tube was inserted between the hydrogen sulphide generation system and the cavity to absorb any water vapours present in the gas mixture. The limit of detection for sulphide was 40 ng when 1 mL sample aliquots were injected and about 10 ng for a 50 µL sample aliquots.

Sulphite can be determined by the release of sulphur dioxide from acidic solutions of the analyte (45). However, since gas evolution is slow, the sulphur dioxide has to be first collected in a liquid nitrogen cold trap before sweeping into the cavity (Figure 4.4).

A specially designed gas generation system has been proposed for the determination of sulphate based on its reduction to hydrogen sulphide (46). The reductant was prepared by heating tin with condensed phosphoric acid (CPA). After cooling, the sample (maximum volume 3 mL) was injected into

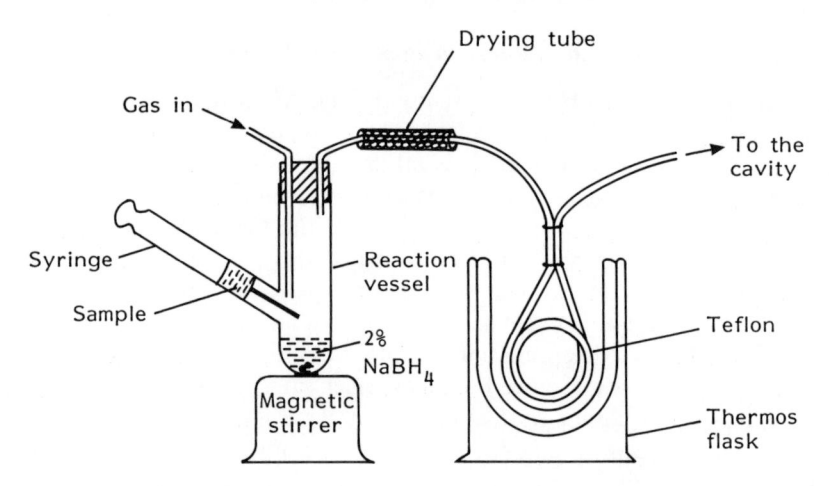

Figure 4.4. Diagram of gas generation system for detection of hydrides. Reprinted with permission from Burguera et al., *CRC Crit. Rev. Anal. Chem.*, **10**, 185 (1980). Copyright CRC Press, Inc., Boca Raton, FL, USA.

a reaction vessel containing 1 mL of Sn–CPA reductant. The gas generation vessel was heated to approximately 198°C, at which temperature sulphate was reduced to hydrogen sulphide, which was then swept into the cavity and determined. Sulphate was determined in the range 40 ng–5 µg with a limit of detection of 15 ng of sulphate.

This procedure has been applied to the determination of sulphate in coal, oil, polyurethane, orchard leaves and distilled water. The same system was also used for the determination of sulphite, thiosulphate and sulphate in admixture by their sequential volatilization. Peroxodisulphate was found to give an identical response to sulphate and thiocyanate gave a similar response to thiosulphate. Shakir (47) has applied the procedure to the determination of chlorpromazine hydrochloride (2–50 mg) in powdered drug tablets. The sulphur in chlorpromazine hydrochloride was oxidized by dichromate in condensed phosphoric acid to sulphate which was then reduced to hydrogen sulphide and determined as above. Since no sample pre-treatment was needed, the procedure required only 10 min to complete.

Sulphite has been determined in simulated soft drinks by injecting the sample solution (10 µL) into a reaction vessel containing 2.5 mL of concentrated phosphoric acid at 72°C and sweeping the sulphur dioxide generated into the cavity (48). By this means, 0.5–6.0 µg of sulphur dioxide was determined with a detection limit of 25 ng of SO_2. Because of the selective evolution of sulphur dioxide from the sample, no other emitting species which might interfere spectrally are introduced into the cavity, and thus the monochromator can be replaced with a blue filter positioned in front of the photomultiplier tube. Calokerinos and Townshend (49) have improved the limit of detection to 5 ng of sulphur and determined 5 ng–4 μg of sulphur as sulphite in 0.5 mL samples using an electronic calibration linearization technique. The procedure was successfully applied to the determination of sulphite stabilized by tetrachloromercurate(II).

Burguera et al. (50) have described a MECA procedure for the determination of total sulphur in fuel oil. The sulphur compounds were reduced to hydrogen sulphide by reacting 1 mL of a sample solution with Devarda's alloy in hydrochloric acid medium at 160°C. The total sulphur recoveries from different sulphur-containing compounds and fuel oils were reported to be about 100%.

Grekas and Calokerinos (51) have reduced organic sulphur compounds by electrolysis and determined the hydrogen sulphide generated. In this way 5–30 μg mL^{-1} of thiosemicarbazide and dithioxamide were determined. Oxygen generated at the anode during electrolysis suppressed the S_2 emission, so to minimize this interference electrolysis was carried out for 30 s and the measurements were taken as soon as electrolysis ceased.

4.2.2.2 Automated Gas Generation Systems

Grekas and Calokerinos (52) have described a continuous flow technique for the determination of 1–50 μg mL^{-1} of sulphite by MECA. The sample was mixed with an excess of orthophosphoric acid in a flow system (Figure 3.15) and the sulphur dioxide evolved was continuously transferred into the cavity for measurement of the S_2 emission. Samples could be analysed at a rate of 24 samples per hour. Most cations suppressed the emission from sulphite by forming stable complexes or insoluble compounds. Calcium(II), barium(II), lead(II) and cadmium(II) interferences were eliminated entirely by masking the cations with EDTA. Those of iron(II), cobalt(II), nickel(II), and copper(II) were reduced significantly by masking with EDTA, but manganese(II) was unaffected. The technique described was applied to the determination of sulphite in soft drinks, white and rosé wines, and sulphur dioxide in air after fixation as disulphitomercurate(II). The method could not be applied to red wines because tannic acid present suppressed the S_2 emission.

Determination of 1–10 μg mL^{-1} of sulphide was also possible using a continuous-flow air-segmented system (53). The effect of nitrite on sulphide determination was eliminated by adding sulphamic acid to the test solutions. Since both sulphide and sulphite produce S_2 emission in the cavity, a method was developed for the determination of both anions in admixture (54). Initially, the total sulphide–sulphite content (A) of the solution was measured. Chloroacetate ions were then added to the solution to convert sulphide into thiodiacetate and, thus, sulphite could be measured without any interference:

$$S^{2-} + 2ClCH_2COO^- \longrightarrow S(CH_2COO)_2 + 2Cl^-$$

Finally, the sulphide–sulphur, which is equivalent to sulphite, was calculated (B) and the concentration of sulphide derived from the difference (A – B).

The method was applicable so long as the final amount of sulphide–sulphur (initial sulphide + sulphide equivalent to initial sulphite) was in the linear range of the calibration graph for sulphide.

Organic sulphur compounds, such as thioacetamide, thiosemicarbazide, thiodiacetic acid and dithiooxamide, have been determined by continuous-flow MECA (55). The compounds were converted to sulphide by alkaline hydrolysis within the delay coil of an air-segmented continuous-flow system. Orthophosphoric acid was then introduced into the stream. The hydrogen sulphide produced was removed from the solution in a debubbler and swept by nitrogen into the cavity for emission intensity measurement. The yield of sulphide depended mainly on the structure of the compound and on the time and temperature of hydrolysis, provided that the concentration of alkali was sufficient. The yields of sulphide for thioacetamide and thiosemicarbazide were 89.6 and 88.8%, respectively, but the corresponding yields for dithioxamide and thiodiacetic acid were 67 and 3.6%. Thus, the limit of detection for the first three compounds was about 1.5 μg mL^{-1} but only 40 μg mL^{-1} for thiodiacetic acid.

The method was also applied to the determination of thiamine (56) and cephalosporins (57) in pharmaceutical preparations. The hydrolyses were carried out at 90°C for 20 min while remaining in the delay coil. The analysis required no sample pre-treatment and 30 samples per hour could be analysed with a relative error of 1–2%. Among the common excipients and other water-soluble vitamins also present in formulations, only ascorbic acid interfered seriously. Cyanocobalamin reduced the emission intensity because of precipitation of sulphide by cobalt ions, but this effect was eliminated by the addition of 0.02 mol L^{-1} EDTA to the stream of sodium hydroxide. In the determination of various cephalosporins, recoveries were quantitative and the results obtained for commercial formulations agreed well with those obtained by standard methods (57).

The applications of automated continuous flow gas generation MECA which have been established so far (Table 4.3), as well as the fact that the gas generation system is capable of reducing chemical interferences and concentrating

Table 4.3. Analytical Characteristics of Various Sulphur-containing Compounds by Continuous Flow Gas Generation MECA

Compound	Gas Evolved	Linear Range (μg ml^{-1})	Ref.
Sulphite	Sulphur dioxide	1.00–50.0	54
Sulphide	Hydrogen sulphide	1.00–10.0	55, 56
Organosulphur compounds	Hydrogen sulphide	2.0–30.0	57
Thiamine	Hydrogen sulphide	20.0–240	58
Cephalosporins	Hydrogen sulphide	10.0–250	59

the analyte, suggest that this technique holds promise for strong future development.

4.2.3 Determination of Sulphur Compounds after Gas and Liquid Chromatographic Separation

Since MECA detection has many features in common with the flame photometric detector (see Chapter 2), a significant amount of research has been devoted to an investigation of whether the cavity can be used as a detector for sulphur and phosphorus compounds after gas chromatographic separation (58).

The separation of isopropyl sulphide and isopropyldisulphide was achieved using a column of 10% silicone gum rubber E-301 on Porapak Q (50–80 mesh). As a further example, thionyl chloride, carbon disulphide and 2-methylthiophene were separated using a column of 10% SE30 on Chromosorb G (80–100 mesh) as shown in Figure 4.5. A simple water-cooled cavity,

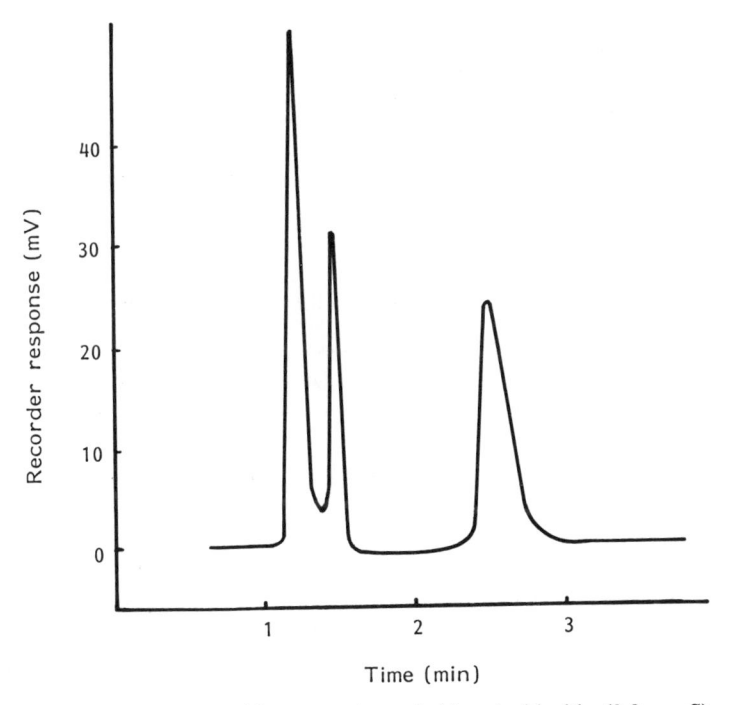

Figure 4.5. Gas chromatographic separation of thionyl chloride (0.2 μg S), carbon disulphide (25 ng S) and 2-methylthiophene (0.3 μg S) (in order of elution) on Chromosorb G column at 190°C with a N_2 flow of 58 mL min^{-1}. Reprinted with permission from Burguera et al., CRC Crit. Rev. Anal. Chem., **10**, 185 (1980). Copyright CRC Press, Inc., Boca Raton, FL, USA.

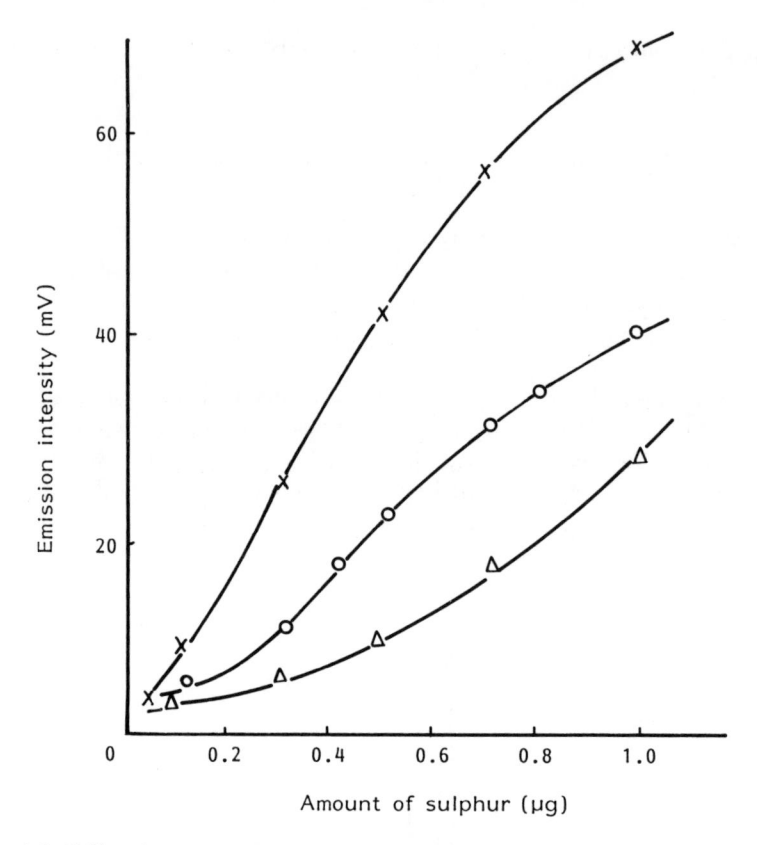

Figure 4.6. Calibration graphs for (×) carbon disulphide, (○) thionyl chloride and (△) 2-methylthiophene with 5 μL samples on the Chromosorb G column at 180°C with 58 mL min-1 nitrogen carrier gas flow. Reprinted with permission from Burguera et al., *CRC Crit. Rev. Anal. Chem.*, **10**, 185 (1980). Copyright CRC Press, Inc., Boca Raton, FL, USA.

was used together with either a hydrogen–nitrogen–air or a hydrogen–nitrogen diffusion flame, the former being slightly more sensitive. No interference was observed from CH radicals which emit at 431.5 nm, because their emission occurred in the flame above the cavity and was not viewed by the detector. Sigmoidal calibration graphs as illustrated in Figure 4.6 were obtained in each case. For isopropyl sulphide, the limit of detection was 15 ng S.

Cope and Townshend (59) noted that MECA could be used to identify the components of eluents from a liquid chromatographic column (see Chapter 3), while Honeyman et al. (60) took the process one stage further and applied the technique to the determination of dixanthogen and sulphur xanthates. The compounds were separated by injecting a 20 μL sample on to a column of either C_{18} reversed-phase or silica gel packing with methanol and water or hexane, respectively, as mobile phase. By injecting 5 μL of the eluate into the cavity and measuring the S_2 emission it was possible to detect 0.013 μg mL^{-1} of analyte.

4.3 SELENIUM AND TELLURIUM COMPOUNDS

Selenium compounds generally give no emission on aspiration of their aqueous solution into a hydrogen–nitrogen diffusion flame or a hydrogen–nitrogen–air flame, while an aqueous tellurium solution gives only a faint blue emission (61). However, when elemental selenium powder was placed in a stainless-steel cavity and then introduced into a hydrogen–nitrogen flame, a weak blue emission was observed. The intensity of the emission was greatly increased by adding air to the flame, as observed with the sulphur compounds (1).

The selenium spectrum consists of several bands superimposed on a continuum, in the range 350–550 nm, with the most intense peak at 411 nm (Figure 4.7). Other selenium compounds such as selenium dioxide and sodium selenate give the same emission spectrum, attributed to Se_2 (2). Tellurium gives rise to a green emission in the cavity, and a blue emission in the flame above.

The green emission spectrum, with a maximum around 500 nm, resembles that obtained previously by Dagnall *et al.* (62) and is probably due to Te_2

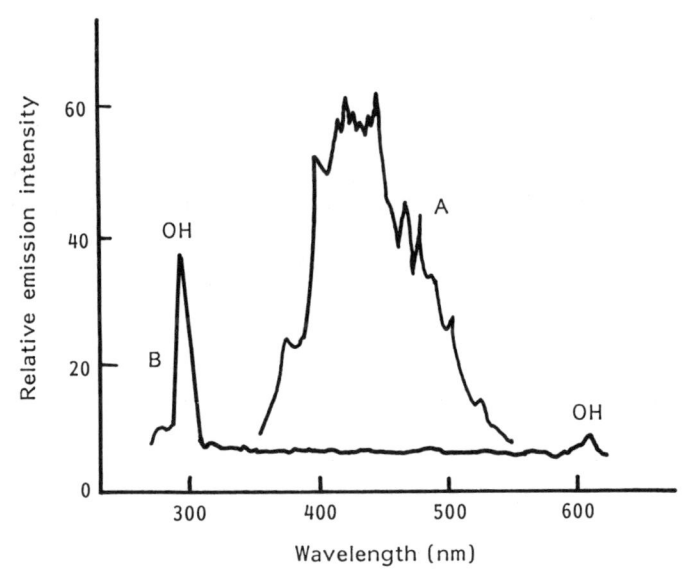

Figure 4.7. Molecular emission spectrum obtained from (A) selenium dioxide in a hydrogen–nitrogen flame (B) flame background. Reproduced by permission of the Royal Society of Chemistry from Ref. 61.

and/or TeO. Tellurium emission is also more intense when a reasonable amount of air is added to the flame.

4.3.1 Determination of Selenium and Tellurium by Conventional MECA

Elemental selenium, selenium dioxide, sodium selenite and sodium selenate each give a dominant Se_2 emitting species when heated in a stainless-steel cavity. The intensity of emission for a given amount of these materials varies, however (63). Selenium powder exhibits two peaks with t_m values equal to 0.5 s and 5.5 s when introduced into a relatively hot flame (5.8 L H_2 min^{-1}, 5.0 L N_2 min^{-1}, 4.0 L air min^{-1}) while selenium dioxide gives only one peak under the same conditions (t_m = 2 s) (Figure 4.8). However, at higher analyte concentrations, selenium dioxide gives a multi-peaked response, typical of the MECA emission from many other compounds in such circumstances (2).

When a stainless-steel cavity is used for tellurium, two peaks are observed (61). The first peak is smaller than the second, which is superimposed on the incandescent emission of the cavity. When a carbon cavity is used, which requires longer residence time to incandesce, the second peak (t_m = 6.3 s) occurs before incandescence, and a third smaller peak is also observed at higher concentrations of tellurium (t_m = 9.3 s) (63). Tellurium dioxide powder also gives three peaks with correspondingly higher t_m values than that of tellurium powder.

4.3.1.1 Inorganic Selenium and Tellurium Compounds

Neither sodium selenite nor sodium selenate (0.5 μg) give any Se_2 emission using flame conditions which are optimal for determining Se in SeO_2. It is thought that at these temperatures these compounds decompose very slowly (63). As with sodium sulphate (2), the selenite and selenate can be treated with phosphoric acid to liberate the non-metal for MECA determination. In addition, phosphoric acid enhances the emission intensity of Se_2 from selenium dioxide by about a factor of 3, without affecting its t_m value. When determining inorganic selenium, therefore, analyte solutions should be prepared 0.1 mol L^{-1} in phosphoric acid.

Tellurium solutions cannot be treated with phosphoric acid because the green HPO emission band (λ_{max} = 526 nm) interferes with the tellurium emission (λ_{max} = 500 nm). Sulphuric acid (0.15 mol L^{-1}) enhances the tellurium emission by a factor of about 8, but the emission peak of sulphuric acid is not completely resolved from that of the tellurium peak.

Organic compounds such as citric acid, ascorbic acid, glucose and tartaric acid enhance the emission from tellurium and selenium. This enhancement has been attributed to a simultaneous production of gases from the organic compound and either selenium or tellurium. A dilution of the analyte vapour

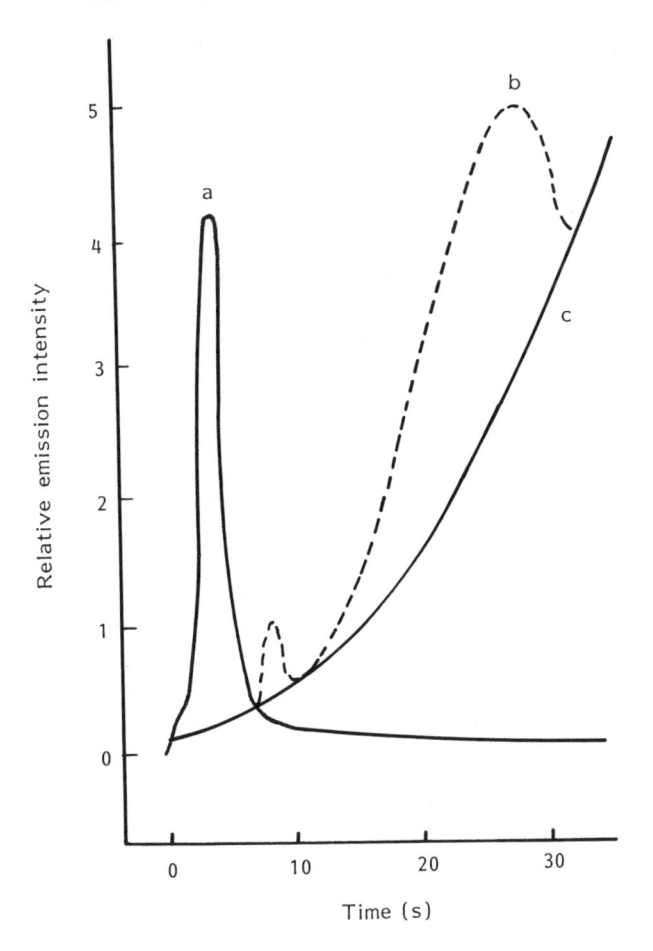

Figure 4.8. Emission profiles from (a) selenous acid (1 μg Se) at 411 nm, (b) telluric acid (10 μg Te) at 500 nm, and (c) incandescence background from a stainless-steel cavity at 500 nm. Reprinted with permission from Burguera et al., CRC Crit. Rev. Anal. Chem., **10**, 185 (1980). Copyright CRC Press, Inc., Boca Raton, FL, USA.

results and prevents the formation of polymeric products which tend to give a multi-peaked MECA response. The exact mechanism has not yet been determined, however (63). Solutions containing tellurium are generally prepared in 0.4 mol L^{-1} citric acid.

In this study (63) the slopes of the log(intensity) vs log(amount of selenium or tellurium) plots were measured at 1.8 ± 0.1, indicating that the emitting species were predominantly Se$_2$ and Te$_2$ rather than their respective oxides. This was also supported by the fact that the emissions were destroyed when a small flow of oxygen was introduced into the cavity. The calibration graphs for selenium and tellurium were S-shaped, with a fairly linear range between 200 and 450 ng for Se and between 5 and 12 μg for Te. The detection limit was 1 μg Te in 5 μL water and 0.6 μg Te in 0.4 mol L^{-1} citric acid. In each instance tellurium was measured as telluric acid.

The co-precipitation technique has been used to determine selenium in selenium sulphide and trace levels of the element in sulphuric acid (61). The method was also used to quantify amounts of selenium in water samples (64). Selenium ions present in 100–250 mL of water samples were co-precipitated with $Fe_2O_3.xH_2O$ at pH 7–8. The precipitate was dissolved in 8 mol L^{-1} hydrochloric acid, after which the selenium(IV) was reduced to the element by hydroxylamine or sulphur dioxide at 70°C. The precipitated elemental selenium was filtered and determined. The analytical range was 0.2 to 15 μg L^{-1} Se; for water samples containing more than 5 μg L^{-1} Se the co-precipitation step was not necessary.

4.3.1.2 Organic Selenium and Tellurium Compounds

For the determination of selenium in organoselenium compounds, individual calibration graphs for each compound must be prepared. Interferences from the organic matrix can be overcome by burning the sample in an oxygen flask and injecting the resulting solution into the MECA cavity. This rapid technique was applied successfully to the determination of small concentrations (0.02 to 0.04%) of selenium in shampoo formulations and also to the determination of selenium in organoselenium compounds at percentage levels (61). When other metal ions are present in commensurate amounts and/or selenium is present in different oxidation states, this technique may not be applicable. If such interfering species are present, selenium can be selectively precipitated as the element using an appropriate reducing agent such as sulphur dioxide or hydroxylamine. The metal ions are filtered through a small glass-fibre filter, which retains particles greater than 1 μm in diameter. The filter containing selenium is then placed in a cavity, introduced into the flame, and the intensity of the selenium emission measured.

4.3.2 Determination of Selenium and Tellurium by Gas Generation Systems

Selenium and tellurium are converted into hydrogen selenide and hydrogen telluride, respectively, by injecting selenium(IV) and tellurium(IV) solutions into a gas generation system containing 1–4% sodium tetrahydroborate(III) (Figure 4.4) (7,65). For MECA determinations, the evolved gases are carried into a 75 cm long (2 mm bore) coiled PTFE tubing immersed in liquid nitrogen. This separates the hydrides from any hydrogen generated in the reaction which would cause an enhanced background emission and which would lead to an increase in noise level.

In the study reported by Belcher et al. (65), a Duralumin MECA cavity (8 mm in diam., 5 mm deep) with a rear gas inlet tube was used. After collecting the hydrides for 2 min, the trap was quickly immersed in water at 90°C to vaporize the hydrides collected which were then swept by nitrogen into the cavity positioned in a hydrogen diffusion flame. The blue Se_2 emission produced by hydrogen selenide was measured at 469 nm to reduce possible spectral interferences from sulphur compounds which reach the cavity by forming hydrogen sulphide or sulphur dioxide.

The calibration graph was linear over the range 20–130 ng of selenium but showed negative deviations at higher concentrations of selenium. The limit of detection was 7 ng of selenium (65).

Although selenium(VI) could not be reduced to hydrogen selenide and could not, therefore, be determined by the gas generation method, it could be reduced quantitatively to selenium(IV) simply by boiling gently for 5 min with 5 mol L^{-1} hydrochloric acid, and then determined. This method is also suitable for determining selenium(IV) and selenium(VI) in admixture.

In the same investigation Belcher et al. (65) showed that 50 µg of Zn(II), Co(II), Ni(II), Cd(II), Fe(III), Mn(II), Al(III), Cr(III), Hg(II), Ba(II), silicate, phosphate, nitrate, sulphate, acetate and oxalate did not interfere with the determination of 0.5 µg of selenium(IV). Other ions suppressed the selenium emission. The maximum tolerable amounts of each for the determination of 0.5 µg of selenium were 25 ng for arsenic(III), 50 ng for antimony(III) and tin(II), 0.2 µg for copper(II) and silver, 0.5 µg for bismuth(III) and germanium(IV) and 20 µg for lead(II). EDTA had no effect on removing these interferences since hydrogen selenide is produced at a low pH, comparable to that used in forming gaseous hydrides of some of the interfering elements (45,66,67,68,69).

Some interferences were minimized by the addition of an excess of tellurium(IV) or by co-precipitating selenium(IV) with lanthanum hydroxide (66). Among the interfering ions, arsenic, antimony, tin, germanium, tellurium and bismuth formed volatile hydrides after reaction with sodium tetrahydroborate and interfered by reacting with selenium in the flame. Other interfering ions were reduced to their metals by sodium tetrahydroborate and formed black suspensions or precipitates which possibly interfered by adsorptive or reactive capture of hydrogen selenide in the reaction vessel.

Spectral interference from arsenic and antimony did not occur because although each gave its characteristic oxide emission, this took place in the flame above the cavity where more oxygen was available. Tin gave a blue SnO emission, mostly above the cavity, and a red SnH emission within the cavity (70) which caused some spectral interference. Germanium gave a red GeH emission (70) in front and above the cavity, which did not interfere spectrally.

The interferences of arsenic and antimony were eliminated by first generating arsine and stibine at lower acidities (solutions 0.1 mol L^{-1} in hydrochloric acid) where hydrogen selenide is not generated, degassing the system and then generating hydrogen selenide by making the solution 2 mol L^{-1} in hydrochloric acid. Tin(II), germanium(IV) and bismuth(III) interferences could not be removed by the same procedure, because of their incomplete conversion to the corresponding hydrides at lower acidities.

Sulphide and sulphite caused spectral interferences from S_2 but this was eliminated by oxidizing the anions to sulphate with hydrogen peroxide, prior to their reduction with sodium tetrahydroborate. The greenish blue emission of even 10 µg of tellurium in the cavity was very weak and, therefore, its spectral interference with the emission of 0.5 µg of selenium was not significant. However, 0.2 to 10 µg tellurium enhanced the emission intensity from 0.5 µg of selenium by about 24% (45,65). The effect of tellurium can be eliminated by adding tellurium to the standard and sample solutions in this range.

The flame containing cavity (see Chapter 3) was also used in the determination of selenium as its hydride (71). This cavity is very successful for the determination of the elements having oxide-based emissions, e.g. boron (71), arsenic, antimony and tin and germanium (72). The emission spectrum from selenium was similar to that obtained by the normal cavity but its sensitivity was lower due to the oxygen-rich nature of the flame within the cavity. The limit of detection was 0.3 µg of selenium.

4.4 CONCLUSIONS

MECA is a simple, versatile and a very effective technique for the determination of sulphur and selenium compounds but the sensitivity for tellurium is not yet high. Determination of mixtures of inorganic and organic compounds is feasible without any sample pre-treatment or separation prior to the analysis, these being requirements which are most necessary in any speciation procedure. Measurements can also be carried out with solid samples, but repeatability and sensitivity are not as good as when samples are in the liquid or gas phase.

Conversion of sulphur, selenium and tellurium to volatile compounds reduces matrix effects and other interferences and improves sensitivity. This aspect of MECA in combination with continuous flow or flow injection automation is the most promising area for future research in the field. Selectivity can be dramatically improved by using the cavity as a detector of a gas or liquid chromatograph.

REFERENCES

1. R. Belcher, S. L. Bogdanski, and A. Townshend, *Anal. Chim. Acta*, **67**, 1 (1973).

2. M. Burguera, S. L. Bogdanski, and A. Townshend, *CRC Crit. Rev. Anal. Chem.*, **10**, 185 (1980).
3. R. Belcher, S. L. Bogdanski, S. A. Ghonaim, and A. Townshend, *Anal. Lett.*, **7**, 133 (1974).
4. K. Nakajima and K. Takada, *Anal. Chim. Acta*, **235**, 413 (1990).
5. S. L. Bogdanski, A. C. Calokerinos, and A. Townshend, *Internat. Lab.*, **14**, 66 (1982).
6. A. C. Calokerinos and T. P. Hadjiioannou, *Anal. Chim. Acta*, **148**, 277 (1983).
7. E. Henden, N. Pourreza, and A. Townshend, *Prog. Anal. Atom. Spectrosc.*, **2**, 355 (1979).
8. R. Belcher, S. L. Bogdanski, D. J. Knowles, and A. Townshend, *Anal. Chim. Acta*, **77**, 53 (1975).
9. S. L. Bogdanski, A. Townshend, and P. Tunon Blanco, *Anal. Chim. Acta*, **131**, 297 (1981).
10. T. J. Cardwell, P. J. Marriott, and D. J. Knowles, *Anal. Chim. Acta*, **121**, 175 (1980).
11. R. Belcher, S. L. Bogdanski, D. J. Knowles, and A. Townshend, *Anal. Chim. Acta*, **79**, 292 (1975).
12. M. Q. Al-Abachi, R. Belcher, S. L. Bogdanski, and A. Townshend, *Anal. Chim. Acta*, **86**, 139 (1976).
13. J. D. Flanagan and R. E. Downie, *Anal. Chem.*, **48**, 2047 (1976).
14. T. S. Al-Ghabsha, S. L. Bogdanski, and A. Townshend,. *Anal. Chim. Acta*, **120**, 383 (1980).
15. R. Belcher, S. L. Bogdanski, I. H. B. Rix, and A. Townshend, *Microchim. Acta*, **11**, 91 (1977).
16. S. Van Vaganen and Q. Fernando, *Anal. Chem.*, **57**, 2743 (1985).
17. A. C. Calokerinos and A. Townshend, *Prog. Anal. Atom. Spectrosc.*, **5**, 63 (1982).
18. T. S. Al-Ghabsha, PhD Thesis, University of Birmingham, 1979.
19. K. Nakajima and T. Takada, *Anal. Chim. Acta*, **199**, 147 (1987).
20. M. Q. Al-Abachi, *Proc. Anal. Div. Chem. Soc*, **14**, 251 (1977).
21. R. Belcher, S. L. Bogdanski, R. A. Sheikh, and A. Townshend, *Analyst.*, **101**, 562 (1976).
22. M. Q. Al-Abachi, PhD Thesis, University of Birmingham, 1977.
23. S. A. Al-Tamrah, PhD Thesis, University of Birmingham, 1978.
24. I. M. A. Shakir, M.Sc. Dissertation, University of Birmingham, 1977.
25. A. Townshend, *Proc. Anal. Div. Chem. Soc.*, **13**, 64 (1976).
26. M. Niezart, PhD Thesis, Liebig University, 1977.
27. S. A. Schubert, J. W. Clayton, and Q. Fernando, *Anal. Chem.*, **51**, 1297 (1979).
28. S. A. Schubert, J. W. Clayton, and Q. Fernando, *Anal. Chem.*, **52**, 963 (1980).
29. J. H. Tzeng and Q. Fernando, *Anal. Chem.*, **54**, 971 (1982).
30. Th. A. Kouimtzis, *Anal. Chim. Acta*, **88**, 303 (1977).
31. Th. A. Kouimtzis, unpublished work.
32. R. Belcher, S. L. Bogdanski, and A. Townshend, US Patent 3,981,585 (1976).
33. M. Q. Al-Abachi, R. Belcher, S. L. Bogdanski, and A. Townshend, *Anal. Chim. Acta*, **92**, 293 (1977).
34. L. Malaprade, *Bull. Soc. Chim. Fr.*, (4th series) **39**, 325 (1926).
35. I. Z. Al-Zamil and A. Townshend, *Anal. Chim. Acta*, **207**, 355 (1988).
36. I. Z. Al-Zamil, Y. A. Hassan, and S. M. Sultan, *Anal Chim Acta*, **233**, 307 (1990).
37. A. Safavi and A. Townshend, *Anal. Chim. Acta*, **128**, 75 (1981).
38. S. A. Al-Tamrah, R. Belcher, S. L. Bogdanski, A. C. Calokerinos and A. Townshend, *Anal. Chim. Acta*, **105**, 433 (1979).
39. I. Z. Al-Zamil, S. M. Sultan, and Y. A. Hassan, *Anal. Chim. Acta*, **239**, 161 (1990).

40. I. H. B. Rix, PhD Thesis, University of Birmingham, 1976.
41. A. Safavi and A. Townshend, *Anal. Chim. Acta*, **164,** 77 (1984).
42. I. H. El-Hag and A. Townshend, *J. Anal. Atom. Spectrosc.*, **1,** 383 (1986).
43. N. P. Evmiridis and A. Townshend, *J. Anal. Atom. Spectrosc.*, **2,** 339 (1987).
44. J. L. Burguera and M. Burguera, *Anal. Chim. Acta*, **157,** 177 (1984).
45. E. Henden, PhD Thesis, University of Birmingham, 1976.
46. I. M. A. Shakir, W. I. Stephen, and A. Townshend, *Analyst.*, **104,** 886 (1979).
47. I. M. Shakir, *Anal. Chim. Acta*, **184,** 295 (1986).
48. S. L. Bogdanski, A. Townshend, and B. Yenigül, *Anal. Chim. Acta*, **115,** 361 (1980).
49. A. C. Calokerinos and A. Townshend, *Fresenius Z. Anal. Chem.*, **311,** 214 (1982).
50. M. Burguera, J. L. Burguera, and J. E. Amaya, *Sci. Total Environ.*, **41,** 163 (1985).
51. N. Grekas and A. C. Calokerinos, *Anal. Chim. Acta*, **202,** 241 (1987).
52. N. Grekas and A. C. Calokerinos, *Analyst.*, **110,** 335 (1985).
53. N. Grekas and A. C. Calokerinos, *Anal. Chim. Acta*, **173,** 311 (1985).
54. N. Grekas and A. C. Calokerinos, *Anal. Chim. Acta*, **225,** 359 (1989).
55. N. Grekas and A. C. Calokerinos, *Anal. Chim. Acta*, **204,** 285 (1988).
56. N. Grekas, A. C. Calokerinos, and T. P. Hadjiioannou, *Analyst.*, **114,** 1283 (1989).
57. N. Grekas and A. C. Calokerinos, *Analyst.*, **115,** 613 (1990).
58. R. Belcher, S. L. Bogdanski, M. Burguera, E. Henden, and A. Townshend, *Anal. Chim. Acta*, **100,** 515 (1978).
59. M. J. Cope and A. Townshend, *Anal. Chim. Acta*, **134,** 93 (1982).
60. T. Honeyman, K. R. Schrieke, and G. Winter, *Anal. Chim. Acta*, **116,** 345 (1980).
61. R. Belcher, Th. A. Kouimtzis, and A. Townshend, *Anal. Chim. Acta*, **68,** 297 (1974).
62. R. M. Dagnall, B. Fleet, and T. H. Risby, *Talanta.*, **18,** 155 (1971).
63. A. Safavi and A. Townshend, *Anal. Chim. Acta*, **142,** 143 (1982).
64. Th. A. Kouimtzis, M. C. Sofoniou, and I. N. Papadoyannis, *Anal. Chim. Acta*, **123,** 315 (1981).
65. R. Belcher, S. L. Bogdanski, E. Henden, and A. Townshend, *Anal. Chim. Acta*, **113,** 13 (1980).
66. J. Azad, G. F. Kirkbright, and R. D. Snook, *Analyst.*, **104,** 232 (1979).
67. R. Belcher, S. L. Bogdanski, E. Henden, and A. Townshend, *Analyst.*, **100,** 522 (1975).
68. E. Henden, *Analyst.*, **107,** 872 (1982).
69. I. Z. Al-Zamil and A. Townshend, *Anal. Chim. Acta*, **209,** 275 (1988).
70. R. W. B. Pearse and A. G. Gaydon, *The Identification of Molecular Spectra,* 3rd ed., Chapman and Hall, London, 1965.
71. S. L. Bogdanski, E. Henden, and A. Townshend, *Anal. Chim. Acta*, **116,** 93 (1980).
72. E. Henden, *Anal. Chim. Acta*, **173,** 89 (1985).

CHAPTER

5

ARSENIC, ANTIMONY, BORON, SILICON, GERMANIUM AND TIN

M. BURGUERA and J. L. BURGUERA

University of Los Andes
Merida, Venezuela

5.1	Introduction	99
5.2	Arsenic and Antimony	100
5.3	Boron	110
5.4	Silicon	114
5.5	Germanium	123
5.6	Tin	125
5.7	Conclusions	128
References		129

5.1 INTRODUCTION

With the development of the 'oxy-cavity' (see Chapter 2) (1), the applicability of the MECA technique has been greatly extended to include almost all the non-metals and metalloids. Arsenic, antimony, boron, silicon, germanium, tin and nitrogen are examples of elements that form oxide-emitting species (see Chapter 2) and may be determined using the oxy-cavity. They cannot be determined by conventional MECA, however, because emission occurs only at the edge of the flame, above the cavity.

Oxide-emitters may be stimulated within the oxy-cavity after generation of hydrides (2–6), silicon tetrafluoride (7,8), methylborate (9,10), ammonia (11) or nitrogen monoxide (12) using a gas generation system. Although gas generation combined with MECA provides a wide range of sensitive and selective applications, its overall applicability has not yet been fully established. This chapter illustrates the capability of gas generation systems for the determination of main group oxide-emitters, which include arsenic, antimony, boron, silicon, germanium and tin.

5.2 ARSENIC AND ANTIMONY

Aspiration of arsenic and antimony solutions into H_2/N_2 diffusion flames produces a faint bluish-white emission of no analytical use. When injected into the MECA cavity, these solutions give emissions outside the cavity which have spectra similar to those obtained on aspiration. They are readily restricted to the confines of the cavity itself by the introduction of a small flow of oxygen (2). Cationic interferences, which have been reported to suppress these signals by the formation of non-volatile salts, may be reduced by using a gas generation system (2,13). In the case of arsine and stibine, these have been generated by the reduction of arsenic(III) and antimony(III) salts with sodium borohydride.

Emission Spectra

An acidified sample solution (1 mL) containing about 1 mg of As(III) or Sb(III) was injected into the reaction tube of the gas generation system containing sodium borohydride. A slow stream of nitrogen mixed with oxygen transported the evolved hydrides into the cavity heated by a H_2/N_2 flame. The emissions obtained were of constant intensity over a period of about 2 min, which allowed the spectra to be recorded. Both elements exhibited broad, banded spectra covering the range 330 to 550 nm, with the most intense peaks occurring at 400 nm for arsenic and at 355 nm for antimony (Figure 5.1). The species responsible for these emission bands are believed to be derived from AsO and SbO, respectively.

Determination of Arsenic and Antimony as Their Hydrides

Preliminary Studies

Experiments carried out to optimize experimental conditions revealed that the most reproducible signals for both elements were obtained when the hydride was generated over a period of 1 min, collected in liquid nitrogen, and then purged by a flow of N_2 and mixed with O_2 prior to being transferred to the cavity (2). A working range of 0.4 to 5 μg for both arsenic and antimony was achieved, with a standard deviation of about 0.14 μg for both elements. The use of helium instead of nitrogen as flame diluent and as a carrier gas doubled the sensitivity for both elements.

Interferences

A detailed study of the interference effect from different species showed that Co(II), Ni(II), Zr(IV), Fe(III), Bi(III), Hg(II), W(VI), Mo(VI), V(V), Se(IV),

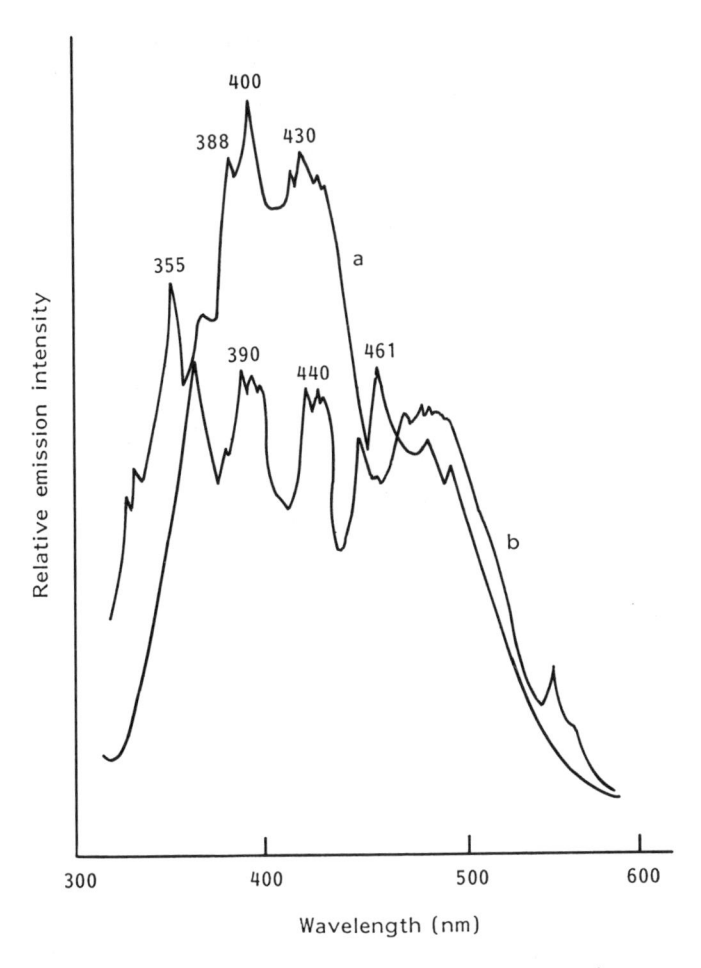

Figure 5.1. Spectra of (a) arsenic, and (b) antimony (H_2 = 3.4 L min^{-1}, N_2 = 5.5 L min^{-1}, O_2 to cavity = 110 mL min^{-1}). Reprinted with permission of Elsevier Science Publishers, BV, from Ref. 2.

Te(VI) and Cd(II) suppressed the evolution of arsine and stibine, while 500 μg mL^{-1} of Al(III), Mn(II), Ba(II), Cr(III), Tl(I), Th(IV), Si(IV), Pb(II), SO_4^{2-}, $C_2O_4^{2-}$, CH_3COO^- and PO_4^{3-} had no effect under the conditions used. Most of the interfering elements gave black precipitates which are believed to be finely divided metal, upon reduction, while zirconium and zinc gave white precipitates due to the corresponding hydroxides. The results in Table 5.1 show that the addition of EDTA entirely eliminates the interferences from some elements. Nevertheless, Bi(III), Cu(II), Hg(II), Te(IV) and Ag(I) still formed black precipitates, Se(IV) and Mo(VI) brown suspensions, while

Table 5.1. Effect of Interfering Ions and of EDTA on Emission Intensities

Ion	Change in Emission Intensity[1], %			
	As Alone	As + EDTA	Sb Alone	Sb + EDTA
Ni^{2+}	−73	2	−75	−5
Zn^{2+}	−58	−4	−31	2
Co^{2+}	−67	−1	−48	−5
Cd^{2+}	−26	−5	−23	0
Fe^{3+}	−45	−3	−40	0
Bi^{3+}	−30	−3	−82	−35
Ag^{+} [2]	−21	−2	−69	−20
Cu^{2+}	−88	−45	−79	−64
Hg^{2+}	−24	−18	−100	−96
In^{3+}	−72	0		
W^{6+}	−44	−2	−59	−50
Mo^{6+}	−18	4	−71	3
V^{5+}			−45	−28
Zr^{4+}			−14	2
Se^{4+}	−59	51	−30	−20
Te^{4+}	−88	−68	−74	−64
$Ti^{3+}, Mn^{2+}, Cr^{3+}$				

Reproduced by permission of The Royal Socielty of Chemistry from Henden (14).
[1] A space indicates no significant interference.
[2] AgCl suspension injected.
Concentrations: arsenic, 0.4 $\mu g\ mL^{-1}$; antimony, 1.0 $\mu g\ mL^{-1}$; foreign ion, 500 $\mu g\ mL^{-1}$; EDTA (where used), 0.05 $mol\ L^{-1}$

W(VI) and V(V) gave blue and violet solutions, respectively. The precipitates from Se(IV) and Te(IV) are believed to be the corresponding elements, which dissolve on being reduced to selenide and telluride. Since the pH of the solution increases to 8.2 soon after injection, the solution is not acidic enough for these two hydrides to be evolved and, therefore, the interference is believed to occur at the hydride formation and evolution step.

The interference from Hg(II) was eliminated by adding 1.6% potassium iodide to the analyte solution, which formed $[HgI_4]^{2-}$. The interference of Ag(I) on the antimony determination was eliminated by removing the silver chloride formed from the test solution. For this purpose, the silver chloride suspension was kept in daylight for 10–15 min, after which it was centrifuged and the supernatant liquid injected. Attempts to eliminate the Cu(II) interference by copper hexacyanoferrate(II) precipitation were unsuccesful, probably due to adsorption or co-precipitation of the analyte ions by the precipitate. However, this precipitation reaction has been applied successfully to the determination of arsenic in copper and its salts with recoveries ranging from 97 to 102% (15). Maximum tolerable amounts of the elements whose

Table 5.2. Maximum Tolerable Amounts of Interfering Ions on the Determination of Arsenic and Antimony

Interfering Ion	Maximum Tolerable Amount[1] ($\mu g\ mL^{-1}$)	
	Arsenic ($0.4\ \mu g\ mL^{-1}$)	Antimony ($1.0\ \mu g\ mL^{-1}$)
Bi^{3+}, V^{5+}, W^{6+}, Cu^{2+}		100
Se^{4+}	250	250
Te^{4+}	4	4

Reproduced by permission of The Royal Society of Chemistry from Henden (14).
[1] A space indicates no significant interference or that the interference could be eliminated by the treatments described.

interferences could not be eliminated by one of the treatments mentioned above are given in Table 5.2. Tin has a strong positive effect on the determination of arsenic and antimony because of spectral overlap.

Gas Chromatographic Separation of Arsine and Stibine

A very interesting way of determining arsenic and antimony simultaneously is based on the gas chromatographic (GC) separation of arsine and stibine, followed by MECA detection of both oxide species at a single wavelength (400 nm) (3). The gas generation system used to produce the hydrides was connected to a PTFE tube (25 cm long × 0.3 cm i.d.) packed with 10% Silicone Gum Rubber E301 on a Porapak Q support (50–80 mesh). Retention times were 61 s for arsine and 245 s for stibine (Figure 5.2), with complete elution of analyte taking place within 362 s. Using a longer column (47 cm) maintained at 9°C and with the carrier gas flowing at 75 mL min^{-1} N$_2$, stannane could also be separated from the other two hydrides (Figure 5.3). Under these conditions the retention times for arsine, stannane and stibine were 132, 251 and 639 s, respectively.

Linear calibration graphs were obtained for arsenic and antimony up to 50 μg mL^{-1}. The standard deviation for the determination of arsenic and antimony depended on how accurately the column temperature was controlled. When the column temperature was controlled to within ±0.5°C, the relative standard deviations ($n = 8$) for the determination of 5 μg of arsenic, by peak height and peak area measurements, were 4.0 and 3.5%, respectively. For the determination of 25 μg of antimony, using peak area measurements, the relative standard deviation was 2.7%. When the column temperature was controlled to within ±0.2°C, the standard deviations decreased to 2.5, 2.9 and 1.8%. The limits of detection (signal = twice background noise) for arsenic and antimony

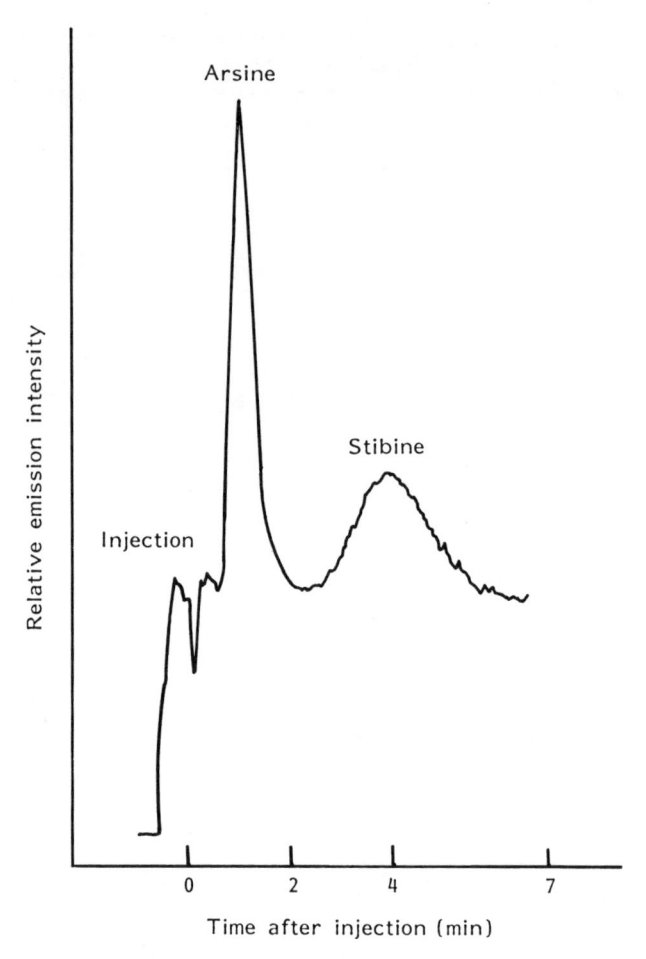

Figure 5.2. Gas chromatographic separation of arsine and stibine (10 μg of As and 50 μg of Sb). Reprinted with permission of Elsevier Science Publishers, BV, from Ref. 3.

were 0.2 and 0.7 μg, respectively. A study of mutual interferences revealed that 5 and 50 μg of antimony do not interfere with the emission from 1 to 10 μg of arsenic, and 50 μg of arsenic do not interfere with the emission from 5 to 50 μg of antimony.

Flame-containing Cavity

The AsO and SbO species can also be generated in a flame-containing cavity (Chapter 3) (16,17). In this configuration, the cavity is connected to a GC column preceded by a hydride generator as described above (3). The shape,

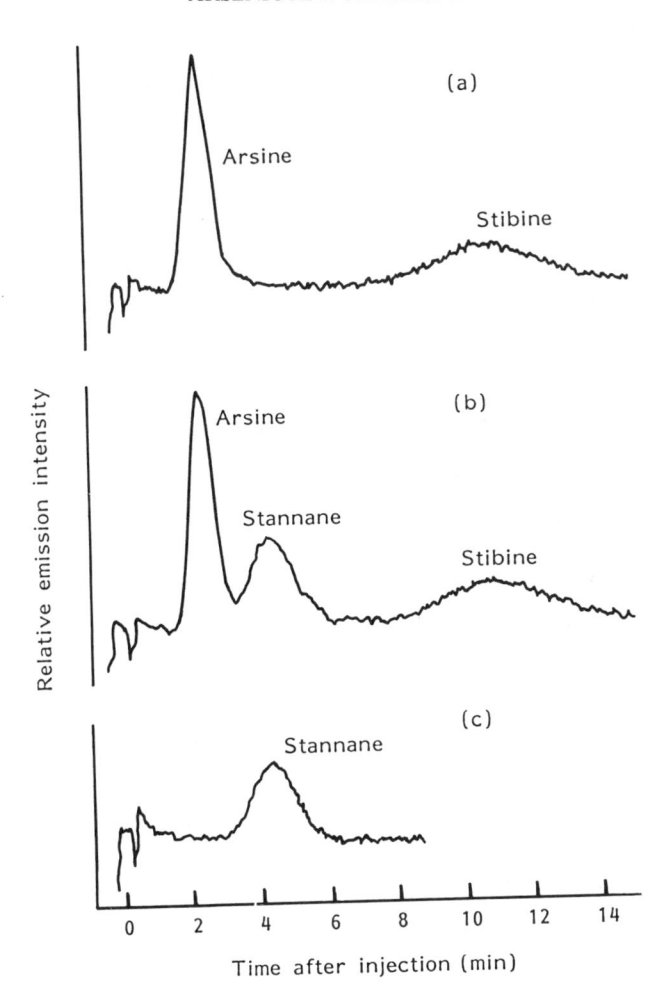

Figure 5.3. Gas chromatographic separation of arsenic (10 μg), antimony (25 μg) and tin (25 μg) hydrides: (a) As and Sb only; (b) As, Sn and Sb; (c) Sn alone; all under the same conditions. Reprinted with permission of Elsevier Science Publishers, BV, from Ref. 2.

volume, temperature and composition of the cavity flame as well as the residence time of the analyte within the cavity are determined by the flow rate of gases reaching it. These are, therefore, optimized for each determination. Successful analyses of arsenic and antimony in binary mixtures as well as the separation of four elements (arsenic, antimony, germanium and tin) in admixture have been made using the flame-containing cavity (Figure 5.4).

Linear calibration graphs were obtained from 0.05 to 3.0 μg of arsenic and 0.1 to 5.0 μg of antimony. The relative standard deviations ($n = 11$) for the

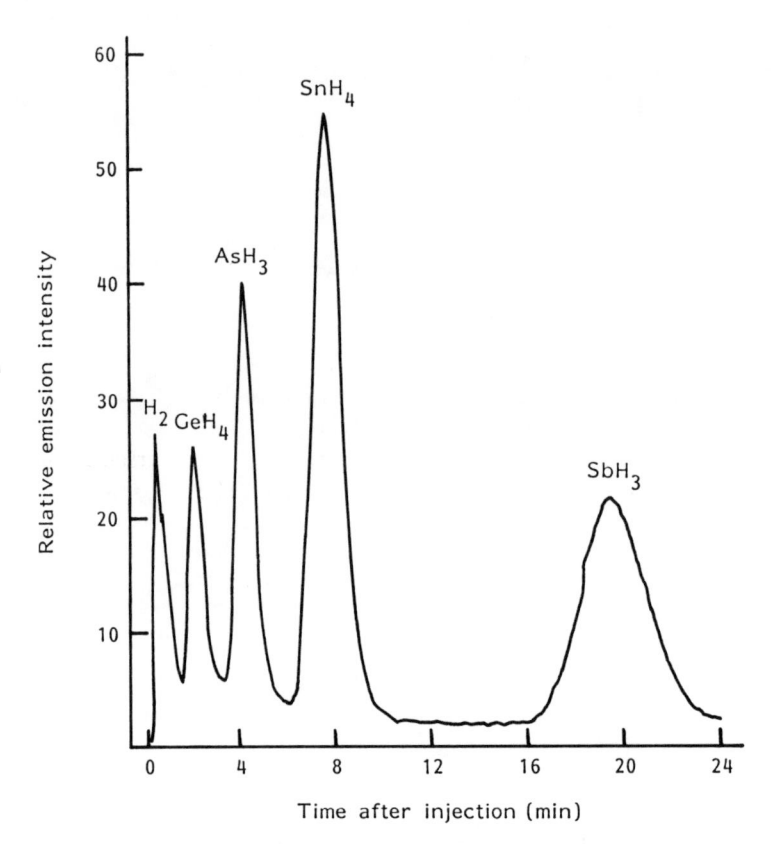

Figure 5.4. Separation of germanium (5 μg), arsenic (3 μg), tin (10 μg) and antimony (20 μg) hydrides under the recommended conditions. Reprinted with permission of Elsevier Science Publishers, BV, from Ref. 17.

determination of 0.4 μg of arsenic and 1.0 μg of antimony in admixture were 3.8% and 4.5%, while the detection limits were 15 ng and 40 ng, respectively.

The use of the GC-separation system did permit some selectivity, but it was not possible to improve the sensitivity without employing a cryogenic trap (3). In this case, the chromatographic column was replaced by a coiled PTFE tube immersed in a Dewar flask containing liquid nitrogen which condensed the hydride and separated it from hydrogen. The cold trap was then removed from the liquid nitrogen and quickly immersed in a water bath at 40°C so that the hydride was swept into the cavity and the emission from arsenic or antimony recorded at 400 nm.

Linear calibration graphs were obtained for up to 2.0 μg of arsenic and 4.0 μg of antimony, with relative standard deviations ($n = 8$) of 4.9 and 3.0% for 1 μg of arsenic and 2 µg of antimony, respectively. These represent a marked improvement in sensitivity over the previous MECA procedure for determining these elements (2).

Flow Injection Analysis

A flow injection (FI) hydride generation MECA analyser has been described by Burguera and Burguera (5). Arsine was generated in the FI system, separated from the liquid phase, and transported with argon to the MECA cavity, where the AsO emission was measured at 400 nm. The configuration of the experimental system used is shown in Figure 3.16. The cavity holder, the circular burner device, the water cooled oxy-cavity and the phase separator used on the FI system have been described earlier (18,19).

The calibration graph obtained by injecting 120 µL of arsenic solutions was rectilinear from 0.1 to 10 µg mL^{-1} of arsenic. The maximum emission intensity, I, expressed in millivolts, increased linearly with arsenic concentration [As] according to the equation $I = -0.12 + 1.52[As]$. The coefficient of variation, r, was calculated as 0.9997 ($n = 7$). The relative standard deviations for the determination of 0.8 and 8.0 µg mL^{-1} of arsenic, were 5.5 and 2.8% ($n = 8$), respectively. The detection limit, considered in this case to be the concentration which generates a signal equal to twice the standard deviation of eight sets of replicate blank determinations, was 0.08 µg mL^{-1}, i.e. 9.6 ng of As.

A variety of likely interfering ions were studied in the FI system under the recommended optimum conditions. An interference was defined as significant if a change of more than two standard deviations in the measurements was observed. Of the ions studied, alkali metals, alkaline earth elements, Al(III) and Cr(III) did not interfere, whereas other ions suppressed the emission intensity. Most of these interferences were effectively eliminated or reduced by the addition of EDTA (Table 5.3), except for Se(IV) and Cu(II). However, the interference effect of these ions was almost eliminated by the addition of 0.20 mol L^{-1} of NaI in the sample carrier solution (Table 5.3).

In order to evaluate the applicability of this method to real samples, total arsenic was determined in NBS standard reference material orchard leaves. Dried sample (1.0 g) was digested using the procedure reported by Liversage and Van Loon (20), after which the final volume was reduced by evaporation to about 4–5 mL. MECA determinations were then performed as indicated above.

The results were in good agreement with the certified values (Table 5.4). The FI-hydride generation analyser has the potential for being applied to the determination of other hydride-forming species which are chemiluminogenic within a MECA cavity. These include Se, Sb, Sn and Ge, elements often found in rocks, sediments and minerals.

Table 5.3. Effect of Foreign Ions, EDTA and Sodium Iodide on Arsenic Emission Intensity

Ion Added	Change in Emission Intensity[1], %			
	As Alone	As + NaI	As + EDTA	As + NaI + EDTA
Ni^{2+}	−50	−43	−3	−2
Ag^{+}[2]	−25	−20	−1	−1
Bi^{3+}	−30	−28	−2	−1
Zn^{2+}	−10	−8	0	0
Te^{4+}	−6	−5	−5	−4
Se^{4+}	−7	−1	−7	−1
Cu^{2+}	−15	−1	−10	−1
Fe^{3+}	−30	−1	0	0
Cr^{3+}	−2	0	0	0
Al^{3+}	−1	0	0	0
Co^{2+}	−34	−33	−1	0

Reproduced by permission of The Royal Society of Chemistry from Burguera and Burguera (5).
[1] Compared with the emission in the absence of interfering ions.
[2] AgCl suspension injected.
Concentrations: arsenic, 1 μg mL^{-1}; all foreign ions at 100 μg mL^{-1}; EDTA and NaI in the sample carrier solution, 0.01 and 0.2 mol L^{-1}, respectively.

Other Procedures for the Determination of Arsenic and Antimony

Precipitation on Fibre Glass Filters

Arsine forms a yellow complex believed to be silver orthoarsenite (Ag_3AsO_3) or a complex composed of silver arsenide and nitrate ($Ag_3As \cdot 3AgNO_3$) when bubbled through a concentrated solution of silver nitrate. The same compound might be formed when the reaction takes place on a filter paper. One possible way of improving the sensitivity and selectivity of the MECA method for determining arsenic would be, therefore, to absorb arsine on a small disc of fibre glass filter containing silver nitrate and to place this in an oxy-cavity for arsenic emission intensity measurements (21).

Table 5.4. Analysis of NBS Orchard Leaves Standard Reference Material

	Arsenic Content (μg g^{-1})	
Sample	Certified	Measured[1]
1	11 ± 2	9.2 ± 0.3
2	14 ± 2	12.0 ± 0.6
3	10 ± 2	9.5 ± 0.4

[1] Four determinations; 1.0 g of dried orchard leaves sample was taken for each analysis.
Reproduced by permisison of The Royal Society of Chemistry from Burguera and Burguera (5).

A conventional hydride generation system was used to generate arsine (1) followed by a drying tube filled with 80-mesh self-indicating anhydrous calcium sulphate (Drierite). This removes traces of water vapour which would otherwise decompose the complex, forming metallic silver and arsenious acid.

$$As_3As.3AgNO_3 + 3H_2O \longrightarrow 3Ag + 3HNO_3 + H_3AsO_3$$

The reaction between the acidic arsenic solution and sodium borohydride in the reaction vessel was exothermic and the gases evolved were hot enough to decompose the silver arsenic complex on the filter and form silver(I) oxide. This effect was reduced by immersing the drying tube in an ice-bath to cool the gaseous mixture before it reached the filter. The filter consisted of a Millipore holder with a dry 5 mm diameter glass paper disc, impregnated with silver nitrate and connected at the end of the drying tube. After the reaction took place, the paper disc was transferred to the oxy-cavity which was then heated in an H_2/N_2 flame for the emission intensity to be measured at 400 nm.

When 0.5 mL of 100 $\mu g\ mL^{-1}$ acidic arsenic solution was injected into 5 mL of 2% sodium borohydride, a canary yellow spot was obtained on the filter which gave the characteristic AsO emission within 8 s of rotation into the flame. Despite taking every experimental precaution to obtain the appropriate complex on the filter paper, seven consecutive injections into fresh borohydride solutions gave a standard deviation of 8% for 50 μg of arsenic. This relatively poor precision was attributed to experimental variables such as an inconsistency in the formation of the silver–arsenic complex on the disc and a variation of the position of the disc within the cavity. Considering the various difficulties encountered in this procedure, the use of the technique for the quantitative determination of arsenic requires further development.

Generation of Arsenic Fluoride

Using optimal conditions for the generation of silicon tetrafluoride (7), arsenic was determined by MECA as its fluoride (21), using the emission at 400 nm. The amount of analyte was varied from 10 to 50 μg, while the amount of fluoride calculated to react stoichiometrically with 50 μg of arsenic to give AsF_3 (40 μg of F) was kept constant. A detection limit of 0.3 μg of As was achieved. Under the same conditions, boron and silicon form their respective fluorides which emit in the same spectral region and therefore constitute interferences.

Formation of Arsenic and Antimony Diethyldithiocarbamates

A substantial improvement in sensitivity was accomplished by converting arsenic and antimony to their diethyldithiocarbamates, extracting the complex into tetrachloromethane and measuring the S_2 emission at 384 nm (22). The

Table 5.5. Analytical Characteristics for Arsenic(III) and Antimony(III) via the S_2 Emission of the Corresponding Diethyldithiocarbamates

Element	As	Sb
Linear range (mg L^{-1})	up to 0.5	up to 1.4
RSD of MECA measurement (%)	2.3	3.5
RSD of complete procedure (%)	4.1	5.0
Detection limit (mg L^{-1})	0.01	0.03

Reprinted with permission of Elsevier Science Publishers, BV, from Ref. 22.

analytical characteristics of the procedure are given in Table 5.5 and indicate that the sensitivities are greater than those obtained by using gas generating systems. The calibration graphs for As and Sb were linear over the ranges studied and not sigmoidal as expected for S_2 emissions (Chapter 2). This is explained by the presence of the $-NCS_2-$ moiety in the ligand, which generates one excited S_2 molecule per molecule of diethyldithiocarbamate decomposed. Other metal ions which were also extracted as their diethyldithiocarbamates interfered with the determination, but this effect was eliminated by the addition of EDTA as masking agent.

5.3 BORON

The earliest observation of molecular emissions from boron compounds was achieved by conventional MECA. The determination of this element was based on the deposition in the cavity of a microvolume of aqueous boron compound, with subsequent heating in an H_2/N_2 flame. Nevertheless, most inorganic boron compounds require high temperatures to decompose, and the boron emission is swamped by the incandescence of the stainless-steel cavity. Since the detector at 518 nm is sensitive to incandescence, boron must be introduced into the cavity in the form of a volatile compound which chemiluminesces before the cavity incandesces.

Emission Spectrum

It is well known that in cool flames (temperatures about 500 K), boron-containing compounds give rise to a green emission, which is attributed to oxide bands (24). They consist of waves of narrow red-degraded bands having maxima at 452.0, 471.0, 493.0, 518.0, 545.0, 580.0, 603.0, 620.0 and 639.0 nm (Figure 5.5).

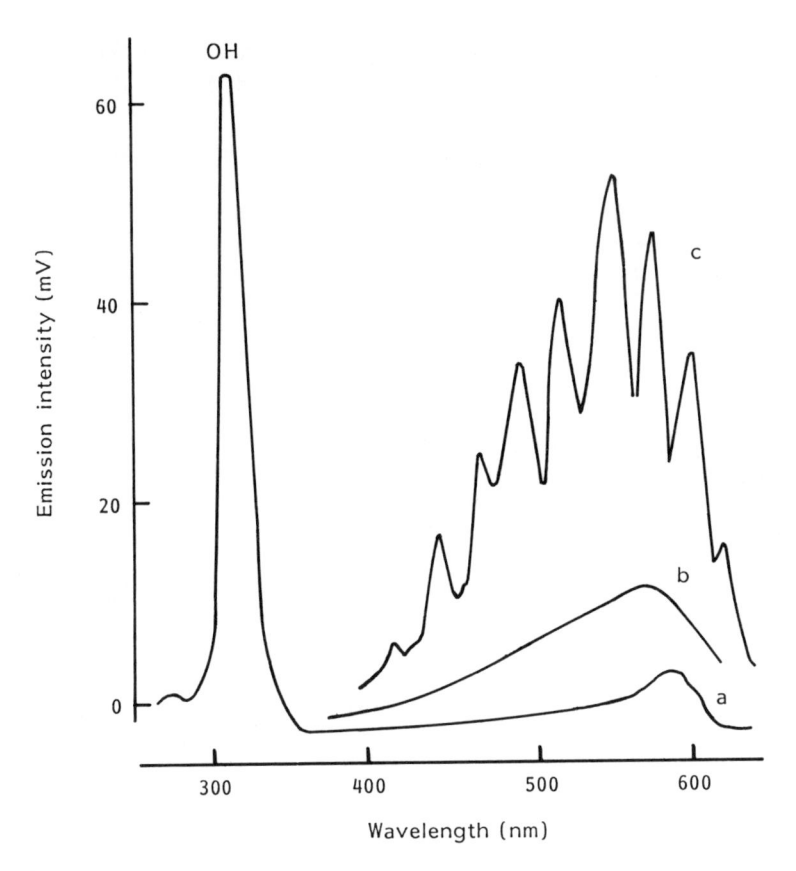

Figure 5.5. Spectra from: (a) flame, (b) methanol blank and (c) boron (Flame: 1.8 L H_2 min^{-1}; 5.5 L N_2 min^{-1}; 80 mL O_2 min^{-1} to cavity; slit 0.2 mm–11 nm). Reprinted with permission of Elsevier Science Publishers, BV, from Ref. 10.

The BO_2 species are believed to be formed by a chemiluminescent reaction such as:

$$OH\cdot + HBO_2\cdot \longrightarrow BO_2 + H_2O$$

which rapidly reaches equilibrium. Belcher et al. (9,10,16) have developed two procedures for boron determination, one based on solvent extraction (9,16) and the other on the evolution of trimethyl borate (9,10). In both cases the BO_2 emission was measured at 518 nm in an oxy-cavity.

Determination of Boron after Solvent Extraction

Boron was chelated with 2-ethylhexane-1,3-diol in acidic solution, and then extracted into methyl isobutylketone (MIBK) (9). Spectral interferences from the blue emission generated by the organic solvent were minimized by evaporating the solvent prior to cavity introduction into the flame. For this reason, the cavity was heated in the flame for 30 s and then cooled for 60 s before the analyte solution was injected. The solution was allowed to stand in the warm cavity for 2.5 min to evaporate the solvent (9).

To obtain an intense BO_2 emission, the flame gas composition and the amount of oxygen added to the cavity, which not only increased the emission intensity from the analyte but also that from the background, was optimized. It was then possible to determine boron in the extract within the range 1 to 16 mg L^{-1}, with a standard deviation of 4% ($n = 8$). The only interference in this procedure was produced by fluoride in amounts greater than 100 mg which suppressed the extraction efficiency by more than 10%.

Using the same extraction procedure, but a flame contained within the cavity, Henden (16) succesfully removed the solvent by first turning on the hydrogen only and igniting it. Introduction of oxygen then produced the BO_2 emission without interference from the solvent. A linear calibration graph was obtained up to 40 mg L^{-1} with a detection limit of 1.16 mg L^{-1} of boron. Although these values are similar to those obtained with the previous procedure, the linear range is at least three times greater.

Determination of Boron after Generation of Trimethyl Borate

A volatilization system based on the generation of trimethyl borate was used as an alternative approach to the extraction procedure for the determination of boron (9). Water was first evaporated and then trimethyl borate was produced by the reaction of boric acid with a mixture of methanol and concentrated sulphuric acid (5:1) at 40°C. The product was carried by a stream of nitrogen into the oxy-cavity. The time required for complete evolution of trimethyl borate under these conditions was about 8 min. A brass cavity was initially used for these experiments but this gave a continuous background emission due to the presence of methanol vapours. The sensitivity was somewhat reduced, therefore, allowing the determination of boron in the range 1.5 to 20 mg L^{-1} with a standard deviation of 3% ($n = 6$). The procedure was considerably improved by various modifications (10). A smaller generator was used (4.5 cm instead of 10 cm high), the reactants were heated to 80°C, and the boron solutions were prepared in methanol instead of water to avoid water evaporation before generation of the gas (Figure 5.6).

The detection limit was found to be governed by the emission originating from methanol, so attempts were made to decrease the blank signal by trapping the methanol vapours either on a molecular sieve or in a cryogenic trap. The first alternative completely eliminated the blank but caused a considerable decrease in the analyte signal, indicating that trimethyl borate was also trapped.

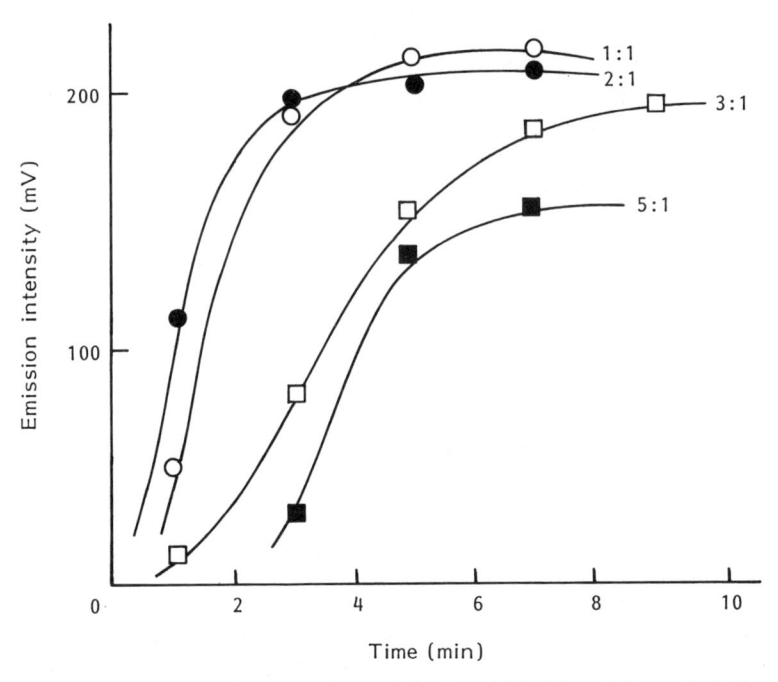

Figure 5.6. Effect of ratio of methanol-to-sulphuric acid (98% w/v) in methylating mixture on emission generated by 20 μg of boron after various reaction times at 80°C (peak height measurement). Reprinted with permission of Elsevier Science Publishers, BV, from Ref. 10.

Complete separation of methanol and trimethyl borate using a cold trap was not expected since their boiling points almost coincide (65 and 68°C, respectively). Nevertheless, when a trap containing solid carbon dioxide suspended in a mixture of equal volumes of chloroform and carbon tetrachloride (−50°C) was used, the blank response decreased by a factor of over five. Much of the methanol seemed to condense in the portion of the tube between the generator and the cold trap, and was not subsequently volatilized with the trimethylborate when the trap was immersed in hot water. This procedure provided sharp MECA peaks recorded over only 5 s with a detection limit of 0.01 mg L^{-1}, while the working range was lowered to 0.25 mg L^{-1} of boron. The method has been succesfully applied to the analysis of steel containing high (15%) to very low (0.001%) concentrations of boron, after its dissolution in sulphuric acid and evaporation to dryness.

Formation of Boron Trifluoride

Preliminary experiments have shown that boron can also be determined by its conversion to boron trifluoride. This is achieved by heating a boron compound in the presence of excess fluoride and concentrated sulphuric acid (21). In this procedure, the same experimental conditions as used in the determination of silicon (7) were employed and the detection limit was only 0.6 µg of boron. Therefore, further study on the optimization of conditions for boron trifluoride generation must be undertaken in order to improve sensitivity.

5.4 SILICON

The H_2/N_2 diffusion flame used in MECA does not provide enough energy to volatilize most inorganic silicon compounds and, therefore, no emission is generated by the conventional technique. However, the emission is stimulated when gaseous silicon tetrafluoride is swept into an oxy-cavity placed in an H_2/N_2 flame (7).

Emission Spectra

The spectrum of the silicon emission in the oxy-cavity was obtained from silicon tetrafluoride, generated by placing 0.5 mL of 1.0 mol L^{-1} hydrochloric acid containing 0.5 mg of silicon as $Na_2SiO_3.5H_2O$ and 2 mg of fluoride as NaF.HF (sodium hydrogen fluoride) in the generator, and adding 1 mL of concentrated sulphuric acid. The gas generated when the system was heated to 135°C was carried to the cavity by a steady stream of nitrogen. A constant white emission which was maintained in the oxy-cavity for at least 2 min showed a broad, featureless emission band with an apparent maximum intensity between 540 and 620 nm (Figure 5.7 (a)).

The only emissions recorded when the spectrum of the flame alone was scanned were from OH radicals at around 306 nm and sodium atoms at 589 nm. The region between 320 and 580 nm was practically free of emission band wavelengths (Figure 5.7 (b)). Hence to avoid any possible effect of sodium emission on the background, all subsequent measurements were made at 580 nm.

The two volatile compounds, silicon tetrachloride and hexamethyldisilazane (HMDS), both produced identical spectra which proves that a silicon-containing species is responsible for the white emission. Of the possible emitting species, SiO and SiF, with dissociation energies 782 and 481 kJ/mol, respectively, SiO was considered the most likely because the spectrum was only observed in an oxy-cavity.

In the absence of oxygen, the behaviour of each of them was different. HMDS generated the white SiO emission above the cavity, while silicon tetrachloride generated several band heads in the region 430 to 570 nm attributed to SiCl species (24) (Figure 5.7 (c)). Yet another silicon-containing compound, silicon

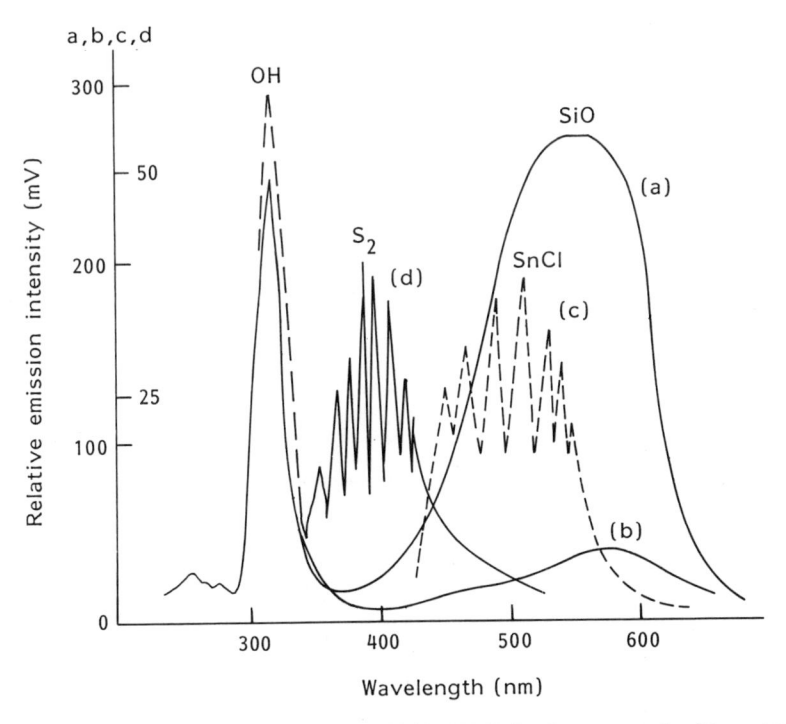

Figure 5.7. Emission spectra from (a) SiF$_4$, SiCl$_4$, HMDS using oxy-cavity (O$_2$ = 120 mL min^{-1}), (b) flame alone and (c, d) SiCl$_4$ and SiF$_4$, respectively, without oxygen supplied to the cavity Reproduced by permission of the author (21).

tetrafluoride produced the blue S$_2$ emission (Figure 5.7 (d)), which suggests that during the period the reaction vessel is heated sulphuric acid vapours and/or sulphur trioxide are formed.

Generation of Silicon Tetrafluoride

The system used for the generation of silicon tetrafluoride is similar to those described in Chapter 2. The generator tube containing an aqueous mixture of sodium meta-silicate and sodium fluoride in hydrochloric acid is placed in a paraffin oil bath maintained at 135°C, so that the following reaction takes place (25–27):

$$SiO_3^{2-} + 6F^- + 8H^+ \longrightarrow H_2SiF_6 + 3H_2O$$

$$H_2SiF_6 \xrightarrow{heat} SiF_4 + 2HF$$

Normally, the evolution of silicon tetrafluoride starts when the reaction mixture reaches the temperature at which sulphuric acid decomposes into

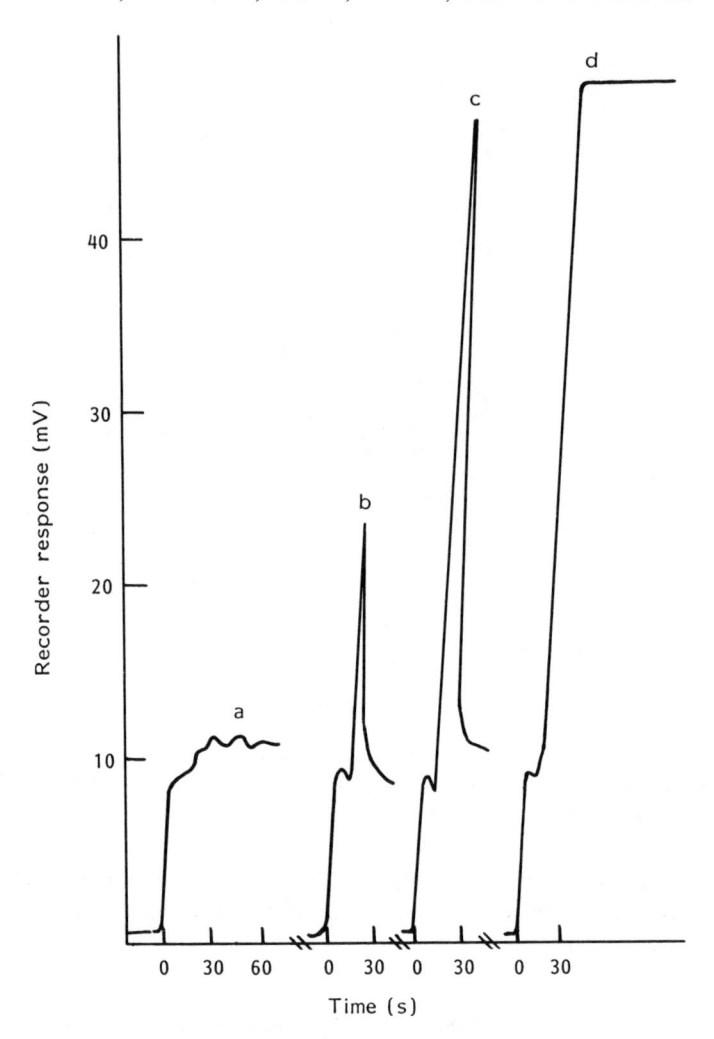

Figure 5.8. Effect of trap saturation on emission intensity: (a) blank with new trap; (b) emission from 20 µg of silicon; (c) same as (b) after 15 experiments; (d) using the same trap after 25 experiments. Reprinted with permission of Elsevier Science Publishers, BV, from Ref. 7.

sulphur trioxide and water (28,29), and is complete within a period of several hours. In this investigation, the evolution of the gas was accelerated by the addition of chloride ions, so that silicon tetrafluoride was evolved within a few seconds. The maximum amount of chloride, which did not show any effect on the blank response (2.0 mg), was used for all experiments. The volatile products formed during the heating period (35 s) were carried to the cavity by a stream of nitrogen.

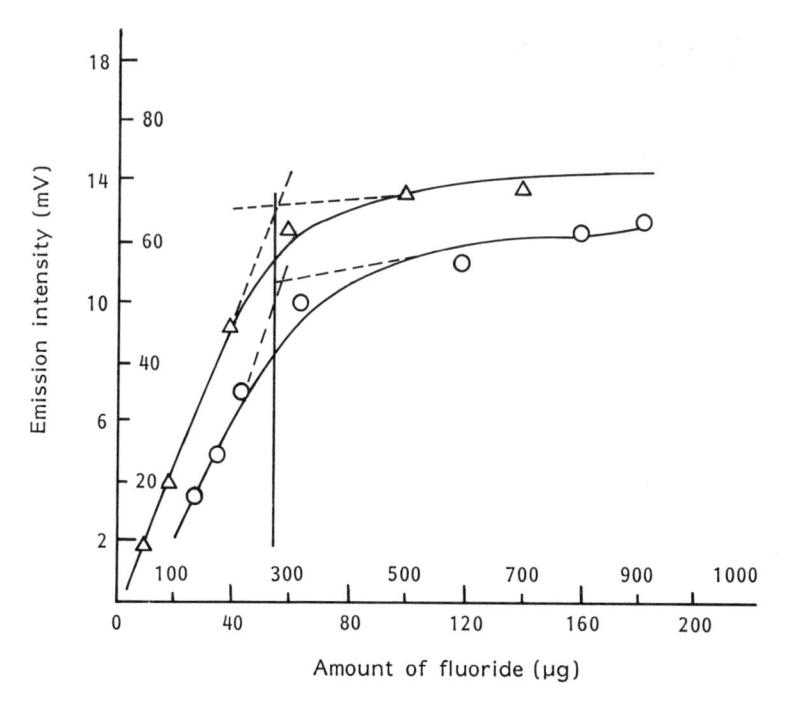

Figure 5.9. Effect of amount of fluoride on the emission from: (△) 20 μg of silicon; (○) 100 μg of silicon. Reprinted with permission of Elsevier Science Publishers, BV, from Ref. 7.

Quantitative recovery of volatilized silicon tetrafluoride is complicated, owing to its strong water affinity. Therefore, the construction and use of the vaporization system must be such that the silicon tetrafluoride along with some hydrochloric and sulphuric acid vapours can be transferred from the sample solution into the cavity without retention by aqueous condensate within the duct. Hence the PTFE tube connecting the generator with the cavity was jacketed with a resistance tape, controlled by a Variac variable resistance which maintained the tube above 100°C. A trap containing 10 mL of 98% w/v sulphuric acid and maintained at room temperature was placed between the heated outer tube and the cavity. This trap absorbed water and sulphuric and hydrofluoric acid vapours, but because it became easily saturated (Figure 5.8), the sulphuric acid was changed every 10–12 consecutive measurements. The amounts of fluoride calculated to react with 20 and 100 μg of silicon to form silicon tetrafluoride are 54 and 271 μg, respectively. The corresponding amounts obtained by experiments and extrapolation (Figure 5.9) were 55 and 275 μg. For subsequent experiments, an excess of fluoride was always used to ensure complete conversion of silicon to silicon tetrafluoride (i.e. 300 μg of fluoride for < 100 μg of Si).

Table 5.6 Effect of Other Ions on the Determination of 20 μg of Silicon

Amount of Ion Added (μg)	Change in Emission Intensity (%) Interfering Ion									
	Pb^{2+}	Hg^{2+}	Co^{2+}	Ni^{2+}	Mn^{2+}	Zn^{2+}	Ba^{2+}	Cr^{3+}	Al^{3+}	Ge^{4+}
10	+3	+4	+6	−2	+6	+5	−2	2.5	0	0
100	+6	+5	−4	+2	+6	+4	+3	+5	+2	−3
1000	+8	+10	+8	+5	+7	+6	+5	+6	+2	−4
Blank	0	0	0	0	0	0	0	0	0	0
	NH_4^+	$C_2O_4^{2-}$	ClO_4^-	PO_4^{3-}	SO_4^{2-}	NO_3^-	BO_2^{3-}	AsO_3^{3-}		
10	+28	−5	+3	−34	+2	+4	+4	+80		
100	+44	−7	+4	−37	+6	+7	+6	>100		
1000	>100	−30	+6	−100	+10	+15	+37	OS[1]		
Blank	OS[1]	2	0	0	0	0	16	OS[1]		

Reprinted with permission of Elsevier Science Publishers, BV, from Ref. 7.
[1] Off scale.

Analytical Characteristics and Interferences

Linear calibration plots were obtained within the range 50–500 μg mL^{-1} of silicon. The relative standard deviations for 5 and 100 μg mL^{-1} of silicon were 12 and 2.3% ($n = 7$), respectively. The detection limit was 1 μg mL^{-1} of silicon.

Several cations and anions were tested for possible interference. Various amounts of interfering ions (50–500 μg mL^{-1}) were added to 100 μg mL^{-1} of silicon. Anions were added as their acids and most of the cations as their nitrates. In each instance, the blank emission was also checked. An interference was defined as significant if the signal from the contaminated solution was different by two standard deviations (i.e. 5%) from the signal obtained from the pure silicon solution. Results tabulated in Table 5.6 show a strong positive interference from arsenic and ammonium, both giving an emission in the absence of silicon. Under the conditions of this experiment, arsenic probably forms arsenic trifluoride which generates the blue arsenic oxide emission in the flame (30) and ammonia produces the whitish NO–O continuum (11). Both these emissions interfere spectrally with the measurement of SiO, but the effect of arsenic and high concentrations of nitrogen can be minimized by using a narrower (0.2 mm) slit, and measuring the emission at 580 nm. Most of the other elements had no significant effect on the determination of silicon.

Determination of Silica in an Iron Ore

In this procedure which was applied to the determination of silica in an iron ore sinter sample supplied by the Bureau of Analysed Samples Ltd., UK, possible interfering elements such as Al, Ti, Mn, Ca, Mg, S, P and V were eliminated by the separation of silicon as insoluble silica (27). The same procedure was also applied to standard silicate solutions in order to construct a calibration graph. Results obtained by MECA were 8.48% SiO_2 for a 0.5 g sample and 8.52% for a 1.0 g sample, which compared well with the certified value of 8.55% SiO_2.

Determination of Silicon Compounds Using MECA as a Gas Chromatographic Detector

Silylation is a reaction widely used in gas chromatography to produce volatile products from hydrogen bonded materials (e.g. carboxylic acids, alcohols, amines and amides), and other non-volatile compounds such as carbohydrates, peptides and steroids. In order to investigate the possibility of determining amino acids by measurement of the SiO emission after silylation, the MECA responses from N,O-bis(trimethysilyl) acetamide (BSA) and hexamethyldisilazane as possible silylating reagents and from acetonitrile (ACN) as a possible solvent were first studied (31).

The column used was a stainless steel tube 190 cm long × 0.4 cm i.d., formed into a coil, 15 cm in diameter and filled with 10% SE30 on Chromosorb G. This

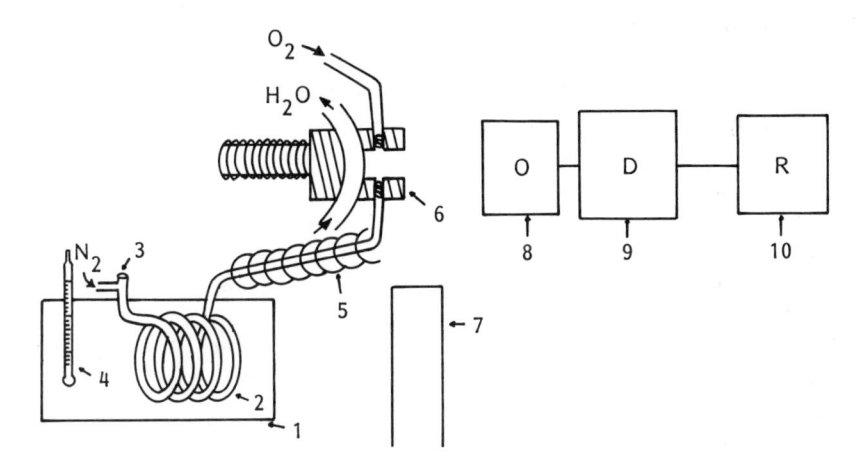

Figure 5.10. Diagram of gas chromatographic system with MECA detection (not to scale) (1. air-circulated oven, 2. column, 3. injection port, 4. thermometer, 5. resistance tape, 6. Duralumin water-cooled oxy-cavity, 7. burner, 8. optical system, 9. detector unit, 10. recorder. Reproduced by permission of the author (21).

Table 5.7. MECA Behaviour of Some Compounds at Different Temperatures of the Column

Compound (abbreviation)	Structure	Retention Time, s at °C		
		150	200	260
Acetonitrile (ACN)	$CH_3-C\equiv N$	78	75	72
Hexamethyldisilazane (HMDS)	$(CH_3)_3Si-N(H)-Si(CH_3)_3$	78, 110	80	75
N,O-Bis(trimethyl-silyl)acetamide (BSA)	$(CH_3)_3Si-N=C(CH_3)-O-Si(CH_3)_3$	78, 110, 170	80, 115	90
Phenyltrimethylsilane (φ-TMS)	$C_6H_5-Si(CH_3)_3$	195	160	127
O-Tolyltrimethylsilane (O-TTMS)	$C_6H_4-Si(CH_3)_3CH_3$	245	198	155

Reproduced by permission of the author (21).

was placed in a controlled-temperature, air circulated oven, and connected to the oxy-cavity in the flame through a stainless-steel tube jacketed with a resistance tape to maintain it at the column temperature and to prevent condensation and/or decomposition of the effluent before it reached the cavity (Figure 5.10).

The SiO emission was measured under conditions already established as being optimal for the measurement of silicon tetrafluoride (7). To diminish the effects from the NO–O continuum (λ_{max} = 500 nm) and the C_2 emission (λ_{max} = 516 nm), all the measurements were carried out at 580 nm. Information in Table 5.7 shows the behaviour of several compounds studied at different column temperatures. At 150°C, HMDS and BSA break down to give two and three peaks, respectively, while at higher temperatures, i.e. 260°C, each compound gives a single peak response.

HMDS is a symmetrical molecule, the two trimethylsilyl groups (TMS) being bonded to one NH group. When these bonds break (Si–N = 435 kJ/mol) and the effluent reaches the cavity, a white emission (NO) appears first, which is followed by the SiO emission.

BSA is a non-symmetrical molecule in which one TMS group is bonded to nitrogen and the other to oxygen (Si–O = 782 kJ/mol). It might be expected that the TMS group bonded to nitrogen would have a shorter retention time than that bonded to oxygen because of its relatively low bond strength. The only experimental proof for this assumption, however, is that the first BSA peak due to silicon has the same retention time as that of the silicon peak for HMDS under the same conditions (Table 5.7).

The presence of different functional groups in amino acid molecules results in the formation of a variety of products. These depend on the strength of the silylation agent employed and on the reaction conditions (32,33). However, since the ease of silylating functional groups is in the order $-OH > -COOH > -SH > -NH_2 > =NH$, possible steps in the formation of trimethylsilyl derivatives of amino acids are:

$$R-CH(NH_2)-C(=O)OH \xrightarrow{\text{TMS donor}} R-CH(NH_2)-C(=O)OSi(CH_3)_3 \xrightarrow{\text{TMS donor}}$$

$$R-CH(NHSi(CH_3)_3)-C(=O)OSi(CH_3)_3 \xrightarrow{\text{TMS donor}} R-CH(N(Si(CH_3)_3)_2)-C(=O)OSi(CH_3)_3$$

Yields for the conversion of aspartic acid and methionine to their TMS-derivatives were established by following the derivatization procedure described by Pierce (34). The amino acids (10–100 µg) were weighed in screw-cap septum glass vials, the containers were tightly closed, and a mixture of 0.5 mL of BSA and 1.5 mL of ACN injected through the septum. These mixtures were shaken for a few seconds and the reaction allowed to progress at 80–100°C until clear solutions were obtained. A 1 µg aliquot of the final mixture was then injected into the chromatographic column at 260°C and the chromatograph recorded (Figure 5.11). The amounts of silicon found to react with BSA to form derivatives were compared with theoretical values for various mono-, di- or tri-TMS derivatives. Results indicate that only di-TMS derivatives were obtained (Table 5.8) and that

Table 5.8. Amount of Silicon Reacting with Amino Acids

Amino Acid Taken (mg)	BSA Consumed (mg Si)						
	Aspartic Acid				Methionine		
	Calculated			Found	Calculated		Found
	Mono-	Di-	Tri-		Mono-	Di-	
20	4.2	8.4	12.6	8.5	3.8	7.5	7.2
50	10.5	21.0	31.6	20.5	9.4	18.8	
70	14.7	29.4	44.2	28.5	13.1	26.3	26.5
100	21.0	42.1	63.1	40.5	18.8	37.5	37.5

Reprinted with permission of Elsevier Science Publishers, BV, from Ref. 31.

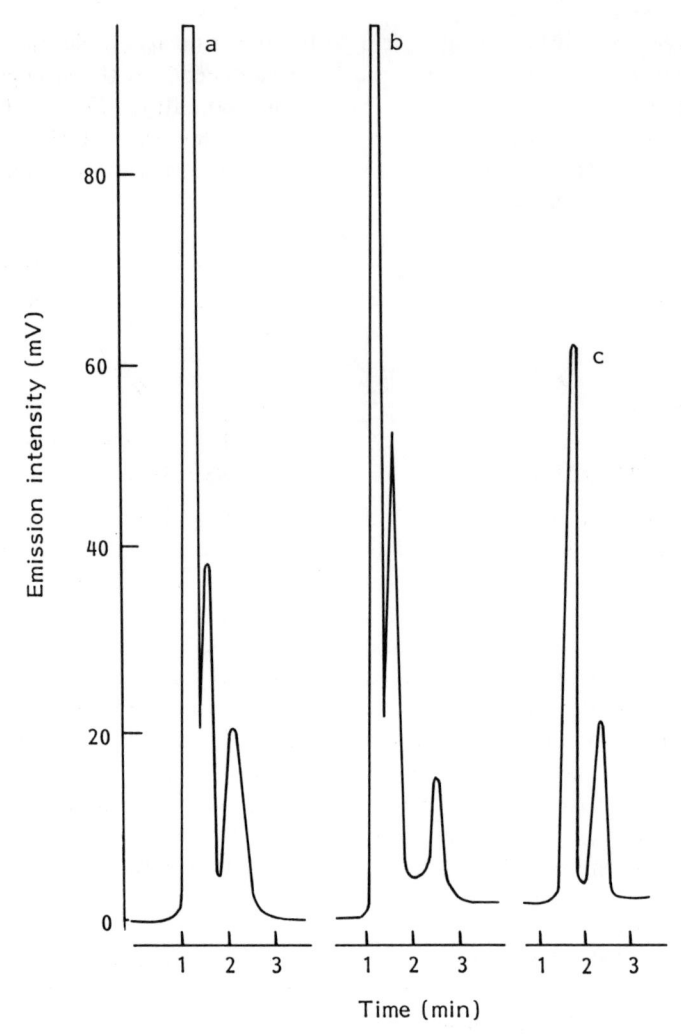

Figure 5.11. Responses from the reaction mixture of: (a) aspartic acid (25 μg), (b) methionine (25 μg) with BSA in ACN and (c) aspartic acid (10 μg) with BSA alone, after separation on the column at 260°C. The last peak in each instance is the amino acid derivative. Reprinted with permission of Elsevier Science Publishers, BV, from Ref. 31.

these are in conformity with those obtained previously by other workers (35,36). The coefficient of variation for the determination of 5 μg of silicon as BSA using ACN as solvent was 22% ($n = 6$) with a detection limit of 0.1 μg of silicon. This is five times more sensitive than when used in conjunction with a volatilization system (7).

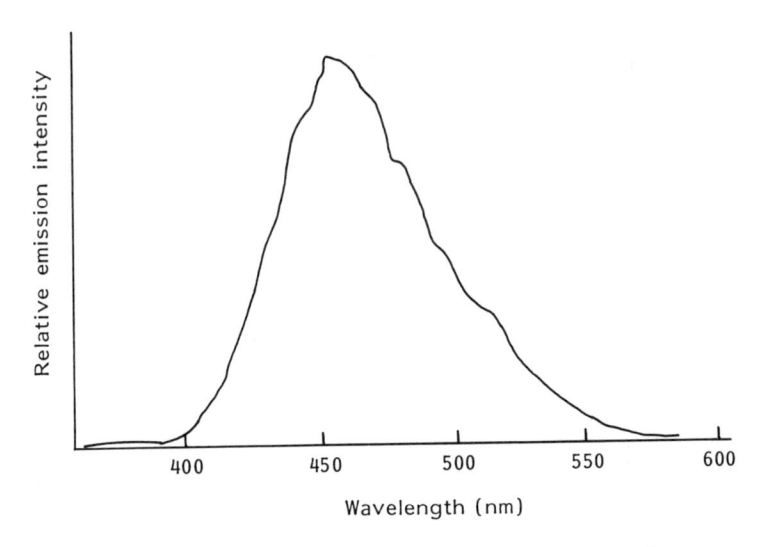

Figure 5.12. Spectrum of GeCl emission obtained from 0.1 μg mL^{-1} germanium in 0.5 mol L^{-1} HCl. Reprinted with permission of Elsevier Science Publishers, BV, from Ref. 4.

5.5 GERMANIUM

No visible emission was observed when an aqueous sodium hydroxide solution containing germanium was injected into any MECA cavity (4). However, germanium solutions prepared in hydrochloric acid gave an intense blue emission when injected into a carbon cavity placed in an H$_2$/N$_2$ flame. The same emission was obtained from germanium solutions prepared in ammonium perchlorate, ammonium chloride or perchloric acid, suggesting that GeCl is the emitting species.

Emission Spectrum

The emission spectrum of GeCl was obtained by conventional MECA using 5 μL aliquots of solution and measuring the resulting emission at different wavelengths. The maximum intensity, obtained by plotting relative emission intensity against wavelength, occurred at 455 nm (Figure 5.12). This wavelength was used subsequently for the determination of germanium by MECA in conventional systems. By using a hydride generation system coupled to an oxy-cavity, germanium generated a broad band spectrum with maximum intensity at 470 nm, attributed to GeO (17) (Figure 5.13).

Determination of Germanium as Germanium Oxide

For the determination of germanium as GeO, a flame-containing cavity

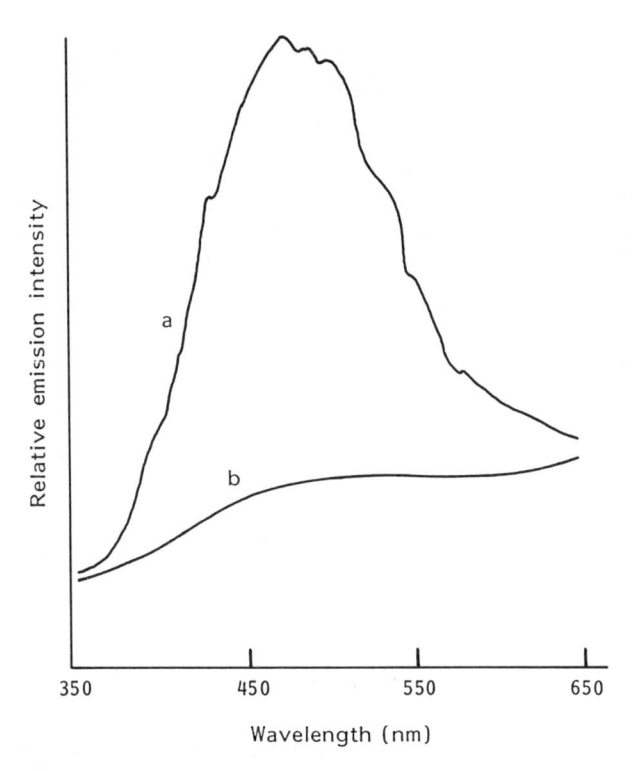

Figure 5.13. Emission spectra: (a) GeH$_4$, (b) flame background (slit = 0.1 nm–3.2 nm at 400 nm). Reprinted with permission of Elsevier Science Publishers, BV, from Ref. 17.

connected to a hydride generation system was used (14,16,17). A linear calibration graph was obtained in the range 0.2 to 10 µg of germanium, with a relative standard deviation of 3.8% for 2 µg of germanium ($n = 8$) and a detection limit of 400 ng of germanium.

The effect of 500 µg mL^{-1} of certain elements on the emission intensity of germanium at a concentration of 2.0 µg mL^{-1} in 0.1 mol L^{-1} hydrochloric acid was studied (14). It was found that Bi(III), Ag(I), Hg(II), W(VI), Mo(VI), V(V), Zr(IV), Ti(III), Mn(II) and Cr(III) had no effect, while Ni(II), Zn(II), Co(II), Cd(II), Fe(III), Cu(II), In(III) and Se(IV) suppressed the emission. Addition of EDTA greatly reduced most of these interferences, probably because it prevented reduction of the interfering element and the subsequent formation of any precipitates. It was not possible, however, to remove selenium and tellurium by any means. The maximum tolerable amounts of these elements on the determination of 2 µg mL^{-1} of germanium were 20 and 50 µg mL^{-1}, respectively.

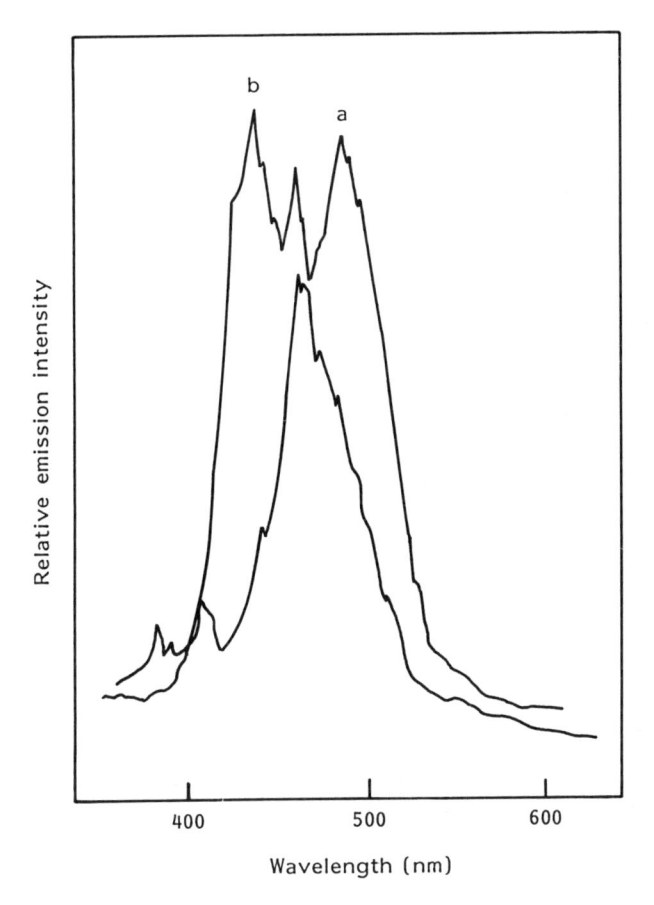

Figure 5.14. The MECA spectra of (a) SnBr and (b) SnCl. Reprinted with permission of The Royal Society of Chemistry from Ref. 30.

5.6 TIN

Tin exhibits an unusual behaviour when its solutions are aspirated into conventional pre-mixed flames, both in emission and atomic absorption spectroscopy. Its sensitivity is higher in cooler flames, suggesting that the temperature is not important in the excitation process and, therefore, chemiluminescence occurs. Any tin compound aspirated into an H_2/N_2 diffusion flame exhibits the same spectrum due to emissions from molecular species located in different regions of the flame (30) (Chapter 2).

Emission Spectra

When tin(II) chloride and tin(II) bromide are injected into a MECA cavity they

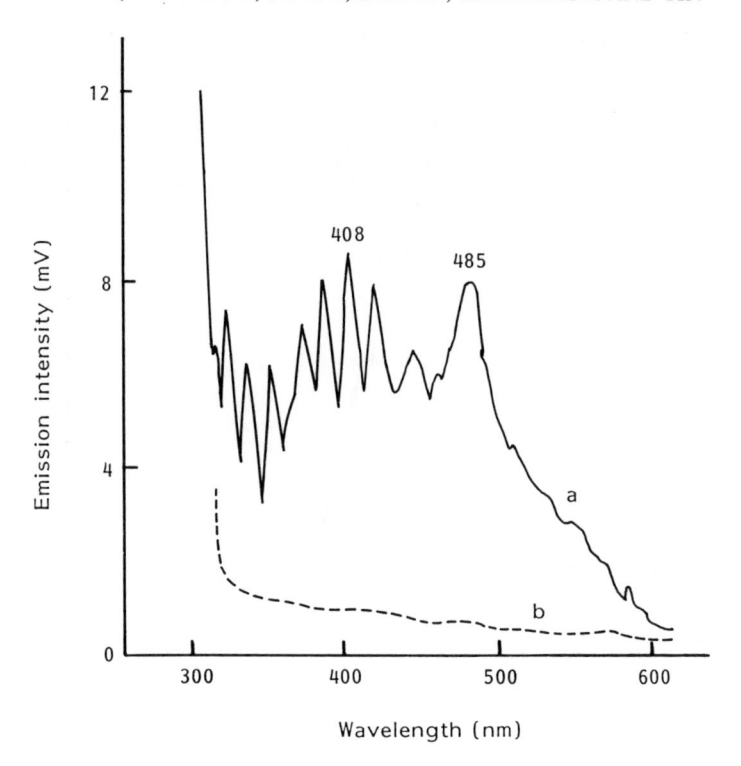

Figure 5.15. Emission spectra of (a) SnO and (b) background. Reprinted with permission of Elsevier Science Publishers, BV, from Ref. 6.

give rise to blue and green emissions, respectively (Chapter 2). The corresponding emitting species are SnCl and SnBr, showing broad emission bands from 400 to 550 nm with maximum peak emission intensities at 440 and 495 nm, respectively (Figure 5.14). The formation of these species indicates that low temperature, low hydroxyl radical concentration, and a specific concentration of hydrogen atoms are necessary. These requirements are fulfilled in the central zone of the flame near the burner head. Both compounds generate a blue emission above the cavity, due to the excited SnO molecule. A small flow of oxygen introduced into the MECA cavity restricts the SnO emission to the region within the cavity, thus permitting measurement of its intensity at 408 or 485 nm (37) (Figure 5.15).

Gas Generation Systems for the Determination of Tin

The formation of the excited SnO molecule allows tin to be determined after its conversion to stannane (SnH_4). This is prepared by injecting a solution of tin(II) in 0.40 mol L^{-1} hydrochloric acid into a generator containing 0.1 g of powdered

Table 5.9 Effect of EDTA on the Tin Emission in the Presence of Some Interfering Ions

Ion	Concentration μg mL^{-1}	Peak Area Supression, %
Co^{2+}	100	5
Ni^{2+}	100	0
Cd^{2+}	10	100
	125[1]	15
Zn^{2+}	2	21
	10	75
	20	88
	100[1]	0

[1] Interference in the absence of EDTA.
Reproduced by permission of The Royal Society of Chemistry from Henden (14).

sodium tetrahydroborate. The stannane is then trapped at 77 K in liquid nitrogen before being released into the cavity by warming the trap to room temperature. Nitrogen gas is used to entrain the stannane, which generates an SnO emission in the oxy-cavity. To prevent hydrolysis of the stannane before it reaches the cavity, it is dried by passing it through a PVC or a glass tube packed with calcium chloride or sulphate granules held between glass wool pads. The calibration graph is linear up to 50 μg of tin(II) and the detection limit is about 80 ng of tin(II) in 1.0 mL of analyte solution.

Stannane production was found to be affected by most metal ions (14,38). EDTA, which has been shown to be effective in reducing interferences in arsenic and antimony determinations based on arsine and stibine generation, caused peak broadening but not peak area reduction in the corresponding tin determination. This was attributed to the slow reduction of tin in the EDTA complex.

Nickel(II) and cobalt(II) completely depressed the tin emission, probably because these ions are reduced by sodium tetrahydroborate to the metal. Metallic cobalt or nickel can then either coprecipitate tin, absorb the hydride formed, catalytically decompose it, or stop its evolution from solution. However, in the presence of EDTA, no interference from cobalt(II) and nickel(II), was observed (Table 5.9). Cadmium(II) and zinc(II), on the other hand, strongly suppressed the SnO emission, probably because these ions form a complex with tin and EDTA which does not decompose readily enough for the tin to be detected.

The effect of EDTA on the rate of hydride generation was eliminated by using the liquid nitrogen cold trap to retain stannane before its transmission to the cavity (6). This procedure also separated the hydride from hydrogen generated in the reaction.

The calibration graph was linear up to *ca* 40 μg mL^{-1} with a detection limit of 80 ng mL^{-1} of tin(II). The relative standard deviation for the determination

Table 5.10 Effect of Various Ions (50 μg mL^{-1}) on the Peak-height Emission Intensity from 10 μg mL^{-1} Tin(II) (Cold Trap Procedure), in the Presence and Absence of EDTA

Ion Added	Intensity Change (%)		Ion Added	Intensity Change (%)	
	Without EDTA	With EDTA		Without EDTA	With EDTA
Cd(II)	−11	−2	Co(II)	−100	5
Fe(III)	−18	−4	Te(IV)	−60	−5
Bi(III)	−10	5	As(III)	100	100
Cu(II)	−40	−3	Sb(III)	100	100
Ni(II)	−100	0			

of 10 μg mL^{-1} of tin(II) was 4% ($n = 7$). By using an aluminium instead of a stainless-steel cavity, it remained cooler during measurement, thereby reducing the thermal decomposition of SnO and improving the sensitivity by about 18% (14). All interferences except those from As(III) and Sb(III) were eliminated (Table 5.10) by employing the cold trap procedure and preparing the tin(II) solutions in 0.02 mol L^{-1} EDTA.

5.7 CONCLUSIONS

The gas generation system interfaced with MECA and described in this chapter involve hydride generation for arsenic, antimony, germanium and tin, trimethyl borate evolution for boron, and silicon tetrafluoride production for silicon. This general procedure, therefore, provides an additional source of selectivity in determining these elements.

Although molecular emissions have been effectively used for the determination of non-metallic elements, methods generally suffer from a lack of sensitivity. They have not, therefore, been used a great deal by analytical chemists.

Molecular emission cavity analysis, however, has been shown to be sensitive, and in this respect has advantages over alternative analytical techniques based on molecular spectroscopy. An important aspect of this work has been the development of a cavity that totally encloses the flame, where the sensitivity is retained but an economy is realized in terms of gas consumption.

However, as opposed to infrared spectroscopy, atomic absorption spectroscopy, inductively coupled plasma spectroscopy, etc., the instrumentation design of MECA has not advanced to the stage where an all-purpose instrument is available for routine analysis. It would seem that this goal can be achieved in the future, though, with an optimization of cavity and holder design, the incorporation of better light collection facilities, and the use of optical filters and non-dispersive detection devices.

CONCLUSIONS

Most of the material devoted to interference elimination in this chapter has been concerned with using chemical reactions prior to actual analytical measurement. A potentially more satisfactory method would entail the incorporation of an on-line separation unit such as an ion-exchange mini-column into the pre-measurement stage of the instrument.

It is exciting to anticipate the vast number of possibilities for further improving the MECA technique. The combination of MECA with FI-automated on-line sample manipulation, or GC/HPLC separations, offers the potential for creating an analytical tool with unrivalled sensitivity and selectivity.

REFERENCES

1. M. Burguera, S. L. Bogdanski, and A. Townshend, *CRC Crit. Rev. Anal. Chem.*, **10**, 185 (1980).
2. R. Belcher, S. L. Bogdanski, S. A. Ghonaim, and A. Townshend, *Anal. Chim. Acta*, **72**, 183 (1974).
3. R. Belcher, S. L. Bogdanski, E. Henden, and A. Townshend, *Anal. Chim. Acta*, **92**, 33 (1977).
4. S. A. Al-Tamrah, I. Z. Al-Zamil, and A. Townshend, *Anal. Chim. Acta*, **143**, 199 (1982).
5. M. Burguera and J. L. Burguera, *Analyst*, **111**, 171 (1986).
6. I. Z. Al-Zamil and A. Townshend, *Anal. Chim. Acta*, **209**, 275 (1988).
7. M. Burguera, S. L. Bogdanski, and A. Townshend, *Anal. Chim. Acta*, **153**, 41 (1983).
8. S. L. Bogdanski, M. Burguera, and A. Townshend, *Anal. Chim. Acta*, **117**, 247 (1980).
9. R. Belcher, S. A. Ghonaim, and A. Townshend, *Anal. Chim. Acta*, **71**, 255 (1974).
10. M. Burguera and A. Townshend, *Anal. Chim. Acta*, **127**, 227 (1981).
11. R. Belcher, S. L. Bogdanski, A. C. Calokerinos, and A. Townshend, *Analyst*, **102**, 220 (1977); **106**, 625 (1981).
12. I. Z. Al-Zamil and A. Townshend, *Anal. Chim. Acta*, **142**, 151 (1982).
13. R. Belcher, S. L. Bogdanski, E. Henden, and A. Townshend, *Analyst*, **100**, 522 (1975).
14. E. Henden, *Analyst*, **107**, 872 (1982).
15. E. Henden. J. Faculty Science Ege University. Series A, **12**, 89 (1989).
16. S. L. Bogdanski, E. Henden, and A. Townshend, *Anal. Chim. Acta*, **116**, 93 (1980).
17. E. Henden, *Anal. Chim. Acta*, **173**, 89 (1985).
18. J. L. Burguera and M. Burguera, *Anal. Chim. Acta*, **157**, 177 (1984).
19. J. L. Burguera, M. Burguera, and D. Flores, *Anal. Chim. Acta*, **170**, 331 (1985).
20. R. R. Liversage and J. C. Van Loon, *Anal. Chim. Acta*, **161**, 275 (1984).
21. M. Burguera, PhD Thesis, University of Birmingham, 1979.
22. A. Safavi and A. Townshend, *Anal. Chim. Acta*, **164**, 77 (1984).
23. I. H. B. Rix, PhD Thesis, University of Birmingham, 1976.
24. A. G. Gaydon, *Dissociation Energies and Spectra of Diatomic Molecules*, Chapman and Hall, London, 1968.
25. I. M. Kolthoff and P. J. Elving, *Treatise on Analytical Chemistry*, Part 1, Vol 1, 2nd ed., John Wiley, New York, 1978.
26. C. Hodzic, *Anal. Chem.*, **38**, 1626 (1966).

27. R. P. Curry and M. G. Mellon, *Anal. Chem.*, **29,** 1632 (1957).
28. R. P. Curry and M. G. Mellon, *Anal. Chem.*, **28,** 1567 (1956).
29. B. D. Hold, *Anal. Chem.*, **32,** 124 (1960).
30. S. A. Ghonaim, *Proc. Soc. Anal. Chem.*, **11,** 138 (1974).
31. R. Belcher, S. L. Bogdanski, M. Burguera, E. Henden, and A. Townshend, *Anal. Chim. Acta*, **100,** 515 (1978).
32. J. P. Hardy and S. L. Kerrin, *Anal. Chem.*, **44,** 1497 (1972).
33. J. Drozd, *J. Chromatogr.*, **113,** 332 (1975).
34. Pierce Chemical Company, General Catalogue, 1966–1977, Method 6, p. 235.
35. K. Bergstrom, J. Guntler, and R. Blomstrand, *Anal. Chem.*, **34,** 74 (1970).
36. K. Bergstrom and J. Guntler, *Acta Chem. Scand.*, **25,** 175 (1971).
37. C. O. Akpofure, R. Belcher, S. L. Bogdanski, and A. Townshend, *Anal. Lett.*, **8,** 921 (1975).
38. A. E. Smith, *Analyst*, **100,** 300 (1975).
39. S. A. Ghonaim, *Proc. Soc. Anal. Chem.*, **11,** 167 (1974).

CHAPTER
6

NITROGEN, PHOSPHORUS, AND CARBON

DAVID A. STILES

Acadia University
Wolfville, Nova Scotia, Canada

and

ALAN TOWNSHEND

The University
Hull, UK

6.1	Introduction		131
6.2	Nitrogen Compounds		132
	6.2.1	General	132
	6.2.2	Indirect Methods	133
	6.2.3	Direct Methods	135
		6.2.3.1 Determination of Ammonia and Ammonium Ions	137
		6.2.3.2 Determination of Nitrite and Nitrate	143
6.3	Phosphorus Compounds		148
	6.3.1	General	148
	6.3.2	Inorganic Phosphorus Determinations	148
	6.3.3	Organic Phosphorus Determinations	154
6.4	Carbon Compounds		162
References			168

6.1 INTRODUCTION

Over the past twenty years, the techniques for the determination of nitrogen and phosphorus in a wide range of compounds have tended to concentrate on the determination of trace and sub-trace levels of these elements quickly, reliably and reproducibly. Methods have therefore been developed to determine nitrogen- and phosphorus-containing compounds at nanogram levels or even lower, in a variety of situations, e.g. forensic, environment, food. Molecular emission cavity analysis (MECA) is one technique that satisfies these criteria in a number of instances. This chapter therefore expands on the development

---

Flame Chemiluminescence Analysis by Molecular Emission Cavity Detection
Edited by D. A. Stiles, A. C. Calokerinos and A. Townshend © 1994 John Wiley & Sons Ltd

of MECA for the determination of both inorganic and organic nitrogen- and phosphorus-containing compounds.

In contrast, the use of this technique for determining carbon-containing compounds has received much less attention. Although emissions attributed to CH and C_2 have been observed from acetaldehyde and other carbon-containing compounds in the MECA cavity, they have not been used as the basis of analytical methods because sensitivity is poor (1). Applications of MECA to the determination of carbon-containing compounds have therefore been restricted to indirect procedures based on the quantification of the more intense S_2 emission. These are described in detail below.

6.2 NITROGEN COMPOUNDS

6.2.1 General

When simple nitrogen-containing compounds are aspirated into flames, they produce several distinct emissions including the narrow NH band (λ_{max} = 336 nm) (2,3), the broad NH_2 band with maxima at 664.2, 630.2, 604.2, 571.3 and 543.6 nm (4,5,6), and an NO−O continuum which extends from 400 nm to the near infrared (6,7). During spectral studies from the MECA cavity using a Vidicon detector, the emission from HNO (analogous to the well-known green HPO emission) was detected in the red region of the spectrum (8). Within the environment of a MECA oxy-cavity, it is likely that the nitrogen emission is due mainly to the NO−O continuum because this species predominates whenever the oxygen concentration is high and the flame temperature is less than 2500°C (9). Experiments with ND_3 have confirmed this assignment because, compared to the spectrum of NH_3, no change in emission wavelength or spectral distribution is observed (10).

Compounds containing both nitrogen and carbon give flame emissions from NO, NH and CN species as well as CH and C_2. The CN emission has been used by several researchers as the basis of analytical methods for determining various organic nitrogen compounds by flame emission spectroscopy (2,11,12).

Observations made with the MECA oxy-cavity have clearly identified the presence of NO, NH and CN emissions, as well as that of the NO−O continuum. Most determinations of nitrogen-containing compounds by MECA involve the direct measurement of a nitrogen emitting species. However, a small number of 'indirect' MECA determinations have been reported where the nitrogen-containing substrate has been converted first to a sulphur-containing species, which is then quantified through the S_2 emission at 384 nm. Both approaches are discussed below as they apply to the determination of nitrogen by MECA.

6.2.2 Indirect Methods

Of the several reactions that involve conversion of a nitrogen-containing compound to one containing sulphur, that of amines with carbon disulphide has been found to be most successful (10). This reaction converts the nitrogen-containing compound to a dithiocarbamic acid by the following reaction.

$$\begin{array}{c} R_1 \\ R_2 \end{array}\!\!NH + CS_2 \longrightarrow S{=}C\!\!\begin{array}{c} SH \\ N \end{array}\!\!\begin{array}{c} R_1 \\ R_2 \end{array}$$

Thus primary and secondary amines are converted to the corresponding acid, while tertiary amines, which contain no displaceable hydrogen, are unreactive. The reaction is quantitative in the presence of an excess of carbon disulphide, and the more readily isolated dithiocarbamic acid salt is obtained by neutralizing the products of this reaction.

Aliphatic primary amines have been observed to react the fastest, while aliphatic secondary amines are, in general, much slower because of steric considerations (13). Aromatic amines react with carbon disulphide at room temperature only if aqueous ammonia or a metal hydroxide is present to force the reaction to completion (14,15).

Certain amino acids also react with carbon disulphide to give the corresponding dithiocarbamino acid. Thus glycine reacts to give dithiocarbamino acetic acid (16). However, since the free acids are unstable, they must be converted to a salt if they are to be isolated.

For MECA determinations, carbon disulphide and the appropriate amine were injected as acetone or n-hexane solutions into a stainless-steel cavity held in a fuel-rich nitrogen-diluted flame. This is an attractive approach because the excess of CS_2 simply evaporates from the cavity at room temperature. The results obtained, however, were unsatisfactory because the S_2 emission intensity decreased with time, most likely because products of the reaction between CS_2 and amines were unstable.

More satisfactory results were obtained when the two reactants were injected into the cavity separately. The general procedure was to inject a solution of carbon disulphide in n-hexane (5 μL) into a cool cavity, followed by an aliquot of the appropriate amine in n-hexane solution (5 μL). The cavity was allowed to stand at room temperature for 1 min and then the S_2 emission intensity was measured after igniting the flame. After 25 s the flame gases were turned off and the cavity allowed to cool by blowing air from a blower over its surface for 2 min. No significant blanks were encountered apart from a deposit of sulphur from carbon disulphide, provided solvents and reactants were of the highest purity. Since the reaction product was formed in maximal amount

Table 6.1 The Determination of Some Aliphatic Amines by the *in situ* MECA–CS$_2$ Method

Amine	Range of Calibration Linearity (μg mL^{-1})	Detection Limit (μg mL^{-1})	Relative Standard Deviation[1] (%)
n-Propylamine	200–600	50	2.3
n-Butylamine	150–500	50	1.9
Diethylamine	20–80	5	1.7
Di-n-propylamine	150–400	50	2.0
Di-n-butylamine	150–350	50	1.7

[1] Calculated from 7 measurements at 100 μg mL^{-1} (for diethylamine, 50 μg mL^{-1}).

within one minute of mixing the reactants, not all CS$_2$ had evaporated after this time, so the CS$_2$ blank obtained one minute after a solution of carbon disulphide alone had been analysed by MECA was subtracted from all S$_2$ emission intensity measurements.

Solutions of n-propylamine, n-butylamine, diethylamine, di-n-propylamine and di-n-butylamine were analysed by the above procedure. The results of these determinations are shown in Table 6.1 where it is clear that the method is most sensitive for the determination of diethylamine. In addition, tertiary aliphatic amines do not interfere in this determination, even when present in six-fold molar excess.

Separate experiments showed that the dithiocarbamate produced by the reaction between carbon disulphide and the aliphatic amine was stable enough to form an identifiable compound which lasted for up to 15 min at room temperature from the time of reactant mixing. Thus, the S$_2$ emission is thought to arise from decomposition of product in the flame. These reaction products are thermally labile, however, so best results were obtained when air was omitted from the flame gas mixture so that the flame was cooler. This experimental condition is different from that described elsewhere (17), where air entrainment in the gas mixture enhanced S$_2$ emission.

In a follow-up study, Al-Tamrah *et al.* (18) modified the carbon disulphide procedure to enable amino acids as well as aliphatic amines to be determined. This method, however, required 3 h for completion of reaction and, therefore, presented severe time limitations.

A more recent investigation by Al-Zamil *et al.* (19) has shown that the amino acids can be determined from the S$_2$ emission obtained from the product of their reaction with 2,4,6-trinitrobenzene-1-sulphonic acid (TNBS). When this reaction solution is buffered at pH 11.7, the reaction is completed within 4 min.

TNBS reacts with amino acids in alkaline solution in the following manner:

$$\text{O}_2\text{N-C}_6\text{H}_2(\text{NO}_2)_2\text{-SO}_3^- + \text{H}_2\text{NCHCOOH} \longrightarrow \text{O}_2\text{N-C}_6\text{H}_2(\text{NO}_2)_2\text{-SO}_2\text{NHCHCOOH} + \text{OH}^-$$
$$\text{(R)} \qquad\qquad\qquad\qquad\qquad\qquad\qquad\qquad \text{(R)}$$

When the solution is acidified, the TNBS-amino acid adduct decomposes, liberating sulphur dioxide:

$$\text{O}_2\text{N-C}_6\text{H}_2(\text{NO}_2)_2\text{-SO}_2\text{NHCHCOOH} \xrightarrow{\text{H}^+} \text{O}_2\text{N-C}_6\text{H}_2(\text{NO}_2)_2\text{-NHCH}_2\text{R} + \text{SO}_2 + \text{CO}_2$$

The sulphur dioxide is then quantified by MECA where the S_2 emission intensity at 384 nm is directly proportional to the amount of amino acid reacting.

Al-Zamil et al. (19) were able to determine glycine, alanine, valine, methionine, serine, threonine and histidine at nanogram levels. Some characteristics of the method are summarized in Table 6.2. Calibration graphs were obtained for each amino acid studied in the range 0 to 4×10^{-4} mol L^{-1}, with acceptable recovery levels in all cases except valine where it was 65%. This anomaly was attributed to steric effects. The limit of detection for all compounds studied (2 × blank noise) was 6×10^{-6} mol L^{-1} (0.06 nmol). One important discovery of this investigation was that reactions between amino acids and TNBS could be completed much faster and without loss of SO_2 if the initial mixture was buffered at the relatively high pH of 11.7 and was chilled in an ice bath before acidification with hydrochloric acid and subsequent MECA detection. This contrasts with previous studies involving spectrophotometry (20,21,22), where it was necessary to work in a much more acidic solution to prevent premature decolourization of the reaction solution and where the reaction time was much slower.

6.2.3 Direct Methods

The indirect determination of nitrogen depends on the fact that the sulphur-containing reactant which converts the nitrogen-containing analyte into an S_2

Table 6.2. Determination of Various Amino Acids by Reaction with TNBS

Parameter	Glycine	Alanine	Valine	Methionine	Serine	Threonine	Histidine
Relative error (%)[1]	−0.1	−5.0	−0.5	−2.0	−2.5	−1.0	−3.5
Relative std deviation (%)[2]	1.84	1.66	2.00	1.80	1.95	2.10	1.75

[1] Using authentic samples in the range 8.0×10^{-5} –2.0×10^{-4} mol L^{-1}.
[2] Six replicates of 2×10^{-4} mol L^{-1}.

emitting species must react quantitatively and quickly for the method to be of any practical use. A more satisfactory approach is to inject the nitrogen-containing compound directly into the MECA cavity and to quantify the amount of nitrogen present by the intensity of the appropriate emission.

6.2.3.1 Determination of Ammonia and Ammonium Ions

When aqueous ammonia is aspirated into a hydrogen-nitrogen flame, the flame becomes white due to the emission of a broad continuum centred at approximately 500 nm (2,23). A similar emission is observed in a stainless-steel MECA cavity if ammonia vapour entrained in a stream of oxygen is transported to it (10). However, this particular experimental design cannot be used for the direct MECA determination of nitrogen because, in the high temperature flame, sodium emissions which come from dust particles or entrained water aerosol and which overlap spectrally with the broad nitrogen continuum interfere.

This problem can be avoided if the MECA cavity is not heated to such high temperatures. A water-cooled aluminium cavity has been designed which eliminates the sodium emission and gives a much more satisfactory environment for observing the nitrogen emission without interference (24). When nitrogen, instead of oxygen, is used to transport the analyte vapour to the cavity, the continuum moves to shorter wavelengths (λ_{max} = ca. 450 nm) and appears blue. In either case, the emission is focused within the confines of the cavity to make the emission more concentrated.

The cavity design used in the first published work on this type of determination was the same as that used in the determinations of boron introduced as methyl borate (25), and arsenic and antimony vaporized as their hydrides (26). It has been described in detail elsewhere in this volume (Chapter 2).

The volatilization apparatus consisted of a 4 mL glass vial containing 0.5–0.6 g of solid sodium hydroxide, screwed on to a PTFE cap fitted with a septum, and inlet and outlet tubes. Oxygen was used as the purge gas and a heating coil was wrapped around the tube leading from the outlet port to prevent water condensing and dissolving ammonia product before it arrived at the cavity. A piece of stainless-steel wire around the oxygen inlet tube within the reaction vessel helped to break up bubbles as they were formed and prevented the outlet port from becoming blocked.

Aqueous solutions of ammonium salts were injected into the reaction vessel as 0.2 mL aliquots. The NO–O emission intensity was measured as the ammonia entered the cavity. Preliminary results based on peak height measurements via a wide spectral slit (68 nm) gave a linear calibration graph up to 10 μg mL^{-1} nitrogen, a limit of detection of 1 μg mL^{-1} nitrogen, and identical results from equal concentrations of nitrogen in ammonium acetate, chloride, and nitrate, as well as aqueous ammonia. No interference was observed from a nine-fold weight excess of Co(II), Cd(II), Cr(III), Mn(II), Zn(II), Ni(II), Pb(II), Cu(II),

Table 6.3. Determination of Ammoniacal Nitrogen in Fertilizers by MECA

Ammoniacal Nitrogen, %

Classical Method[1]	MECA
0.88	0.88
2.03	2.12
4.7	4.6
4.7	4.5
5.0	4.8
5.4	5.5
6.4	6.3
8.4	8.1
9.4	9.4
13.1	13.1

[1] Distillation-titrimetry.

Hg(II), Fe(II), Ag(I), SO_4^{2-}, PO_4^{3-}, NO_3^-, and urea. The method was applied to the determination of ammoniacal nitrogen in fertilizers with excellent results (Table 6.3).

Further investigations along the same lines (27) revealed that although the experimental design was sound, improvements could be made which would increase the flexibility of measurement and allow more experimental parameters to be studied. For instance, the cooled cavity was modified so that oxygen and nitrogen could be introduced through two separate side ducts, and thus avoid potential problems with flashback from hydrogen or other combustible gases formed in the reaction vessel (Figure 3.5 (c)).

The cavity cooling system was changed so that incoming water was directed to the area immediately behind the rear of the cavity to give better control over the cavity temperature (Figure 3.5 (c)). For best results, nitrogen (75 mL min^{-1}) was used as the carrier gas, but oxygen (110 mL min^{-1}) was also directed into the cavity through the second entry port. The temperature inside the cavity was maintained high enough for appreciable NO–O emission to be stimulated and the ammoniacal nitrogen determined.

The degree of water cooling was also found to be an important factor in determining the NO–O emission intensity. At comparatively high water flow rates (100 mL min^{-1}), the cavity became so cool that water condensed from the flame onto the surface of the cavity. Experiments conducted to measure the temperature within the cavity showed that a relatively hot cavity temperature enhanced the ammoniacal-nitrogen emission, so a water flow rate of 10 mL min^{-1} was used throughout.

The amount of sodium hydroxide used to convert ammonium salts into ammonia, and the amount of analyte injected into the reaction vessel, were

Table 6.4. Determination of Ammonium-nitrogen in Contaminated River-water Samples

The values are given in $\mu g\ mL^{-1}$

Sample	1	2	3	4	5	6
MECA	25	26	26	35	35	36
Quoted[1]	28	26	28	35	35	35

[1] These values were provided by the Severn-Trent Water Authority, Coventry, and were determined by spectrophotometric means.

also investigated more thoroughly. Different volumes of aqueous ammonium chloride solution, each containing 50 μg of ammonia, were injected into the reaction vessel. When this was 0.4 mL or more, the sodium hydroxide which remained in the reactor after all the ammonia had been generated became syrupy. Emission intensity peaks were broad and unsatisfactory for measurement purposes, probably because any ammonia formed tended to dissolve in the liquid present in the reaction vial or was prevented from diffusing rapidly into the carrier gas stream. When the volume of injected solution was less than 0.05 mL, the purge gas inlet tube was not immersed. This condition led to incomplete transfer of any ammonia formed to the cavity. Hence in all experiments undertaken subsequently, the volume of injected solution used was 0.2 mL.

The results reported in this study (27) confirmed the utility of the method for determining ammoniacal nitrogen. A single calibration graph, linear within the range 10 $\mu g\ mL^{-1}$ to percentage level, could be used for all ammonium salts analysed; the limit of detection using peak heights (signal to noise ratio = 2) was 1 $\mu g\ mL^{-1}$ nitrogen; the coefficient of variation for 10 measurements of 100 $\mu g\ mL^{-1}$ nitrogen was 2%.

Applications of this method were extended from measuring levels of ammoniacal nitrogen in fertilizers to those present in contaminated river waters, effluents and coke-oven liquors. For river water samples, the MECA method showed excellent correspondence with results obtained spectrophotometrically (Table 6.4). Results for ammoniacal liquors are given in Table 6.5. Because these samples contained foaming agents, which made it difficult to prevent the vapour ducts from becoming blocked up, a second trap had to be incorporated into the ammonia delivery system.

A further examination of ammoniacal nitrogen determination has been made by Celik and Henden (28) who used a MECA system where the entire heating flame was contained within the cavity (Chapter 2). This made it possible to use smaller quantities of gases and thus make the determinations less costly. In addition detection limits were improved considerably over earlier studies (24,27).

Celik and Henden utilized a stainless-steel MECA cavity which had been designed for a previous investigation by Henden (29). This was modified by

Table 6.5. Determination of Ammonium-nitrogen in Effluents and Coke-oven Liquors

The values given are in $\mu g\ mL^{-1}$

Sample	1^1	2^1	3^2	4^2	5^2	6^1	7^2	8^2
MECA	130	280	350	880	1025	1500	2400	4500
Quoted	140	264	331	911	1066	1443	2231	4500

[1] Coke-oven liquors provided by the British Carbonisation Research Association.
[2] Effluent liquors provided by the British Gas Corporation.

incorporating a copper tube into the cavity assembly for the supply of water coolant (Figure 3.6). The reaction vessel was also modified from that used by Belcher et al. (26), as shown in Figure 6.1. In this instance the reaction vessel was bent to allow direct injection of sample through a septum placed in its side wall. The nitrogen transfer line included a three-way valve attached to the outlet tube from the reaction vessel and to the direct line from the nitrogen storage cylinder to the cavity. It was therefore possible to purge the complete system of air before any ammonia was produced.

Consistent with all similar previous investigations, pure nitrogen admitted to the cavity in the presence of air gave no emission (30), but when ammonia vapour was entrained in the same gaseous mixture a broad continuum (λ_{max} = 640 nm) was observed. Although this band has a longer wavelength of maximum emission than that already attributed to NO−O (λ_{max} = 500 nm) (26, 29), it was assigned to the same emitting species. The discrepancy was thought to be caused by differences in flame temperature and composition.

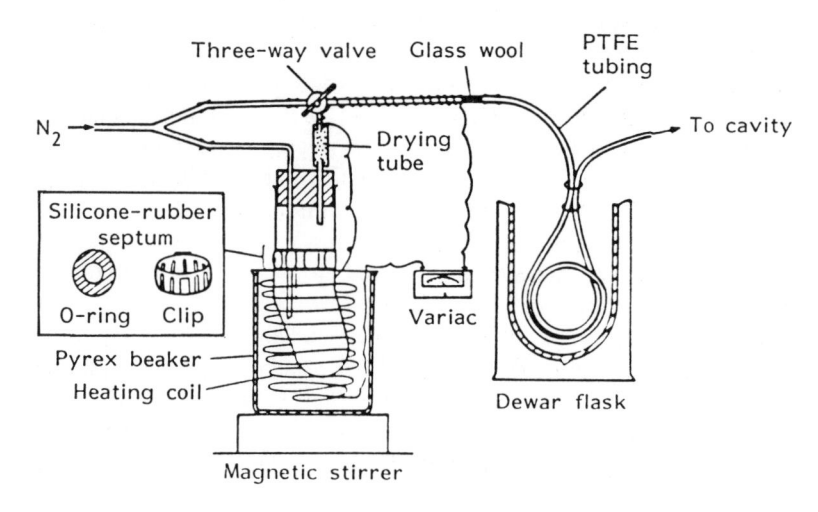

Figure 6.1. Volatilization system for the determination of ammonium-, nitrite- and nitrate-nitrogen.

The overall objective of the study was to improve detection limits for the MECA determination of ammoniacal nitrogen. This was achieved in two ways. The first of these methods was based on the conventional reaction of sodium hydroxide with analyte, followed by transfer of the ammonia produced to the cavity. However, in this particular experiment, the whole system was first purged with nitrogen gas. Then the three-way stopcock was set so that nitrogen gas by-passed the reaction vessel. Analyte was added to the sodium hydroxide in the reaction vial and allowed to react for two minutes at 85–90°C. The three-way stopcock was then adjusted so that the nitrogen flowed through the reaction vessel, allowing the ammonia generated to be swept into the cavity. The amount of ammoniacal nitrogen in the analyte was determined by measuring the peak height of the NO–O emission response at 640 nm. The limit of detection was 1.5 μg mL^{-1} nitrogen with a linear calibration range from 5.0 to 100 μg mL^{-1}. The relative standard deviations (seven replicates) for the determination of 20 and 40 μg mL^{-1} ammoniacal nitrogen were 5.9% and 4.9%, respectively.

In the second procedure, a cold trap, consisting of a coil of PTFE located between the three-way stopcock and the cavity, was immersed in liquid nitrogen contained in a Dewar flask. A substantially larger volume of analyte solution (5 mL) was injected into the reaction vessel which was allowed to stand without applying external heat for 4 min. The ammonia produced was then swept into the Dewar flask through a drying tube containing dry sodium hydroxide which prevented any water from the reaction vessel interfering. The cold trap was then removed from the liquid nitrogen and heated in a water bath at 90–95°C so that the ammonia was transferred to the MECA cavity. The NO–O emission intensity was measured as before. The limit of detection for this method was 0.01 μg mL^{-1} nitrogen. Linear calibration graphs were obtained over the range 0.05–1.0 μg mL^{-1} nitrogen with a relative standard deviation of 4.1% for 0.5 μg nitrogen (seven replicates). At concentrations of ammoniacal nitrogen higher than 1.0 μg mL^{-1}, the calibration graph started to level off. This is probably due to self absorption of the emission at the higher concentrations of analyte.

Interesting further applications of MECA to the determination of nitrogen in agricultural and environmental samples has been provided by Shakir and co-workers (31,32,33). Ammoniacal nitrogen in soil samples was extracted according to the method of Hesse (34). Samples of 1 g capacity were first moistened with water and then allowed to stand for 15 min in a solution of potassium chloride (2 mol L^{-1}; 10 mL). These were then filtered through filter paper saturated with potassium chloride to prevent any ammonium ions being co-extracted during this procedure. Portions of each filtrate (1 mL) were then placed in 4 mL dry glass reaction vessels to which were added seven pellets of solid potassium hydroxide. The reaction vial was connected immediately to an air supply which transferred the ammonia produced into the water-cooled

MECA oxy-cavity. Emission intensities were measured as peak heights at a nominal wavelength of 500 nm. Although satisfactory measurements could be made without heating the reaction vessel, the best results were obtained when it was heated to 175°C by placing it in a heating block.

This procedure appears to show promise for determining ammoniacal nitrogen in soils since it can be accomplished in 1 min. Shakir *et al.* (31) investigated a total of 13 soil samples with ammoniacal nitrogen concentrations ranging from 17.3 to 51.7 mg kg^{-1}. The coefficients of variation and standard deviations were well within the acceptable norms for standard analytical procedures.

A further investigation by Shakir and co-workers used MECA to determine gaseous ammonia and ammonium ion levels in artesian wells in the Arbil Governate of Iraq. For the ammonia determinations, air from these wells was pumped at 300 mL min^{-1} through an absorption bottle containing hydrochloric acid (3 mol L^{-1}) (Figure 6.2). The ammonium ions formed were added to solid potassium hydroxide and nitrogen was determined from the subsequent NO–O emission. In the case of water samples, they were added as aliquots (1 mL) directly to potassium hydroxide pellets.

Apart from studying the levels of ammonia and ammonium ions present in each of the artesian wells, the other important aspect of the investigation was confirmation that gaseous carbon dioxide, which would be produced from the reaction of carbonate-contaminated potassium hydroxide with acid, does not interfere in either of the determinations.

The most recent investigation by Shakir *et al.* (33), concerning the determination of total nitrogen in tobacco leaves, shows that MECA can be used as a viable alternative to more lengthy titrimetric procedures. Using a Hoskins distillation apparatus, a small portion (0.5 mL) of the acid digest solution was diluted to 1.5 mL with distilled water and then allowed to react with potassium

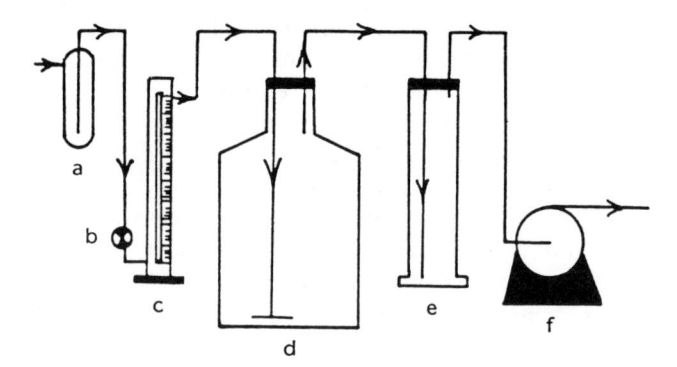

Figure 6.2. Schematic diagram showing sampling of air from the atmosphere before MECA determination: (a) inlet guard filled with glass wool (20 mL); (b) needle valve (one way); (c) air flow meter (mL min^{-1}); (d) adsorption bottle (200 mL); (e) guard tube (20 mL); (f) air pump (FRACMO Fractional H. P. Motors Ltd, Hastings, UK).

hydroxide in the reaction vessel. The remainder of the apparatus was the same as that described above, so that nitrogen levels were determined from the intensity of the NO–O emission. The study showed that MECA has the advantage in Kjeldahl analyses of being faster and more economical than those based on titrimetry.

6.2.3.2 Determination of Nitrite and Nitrate

The first indication that the MECA technique could be used for the determination of nitrogen in oxidized species was given by Belcher et al. (24). They suggested that replacement of sodium hydroxide pellets by Devarda's alloy in the reaction vessel of the ammonia generation MECA system would provide the necessary reductant to convert nitrate to ammonia. Subsequently, this method was investigated in more detail (27).

Devarda's alloy, which contains copper, aluminium and zinc, reduces nitrates to ammonia in basic solution, liberating hydrogen in the process (35). When this hydrogen is entrained in the stream of nitrogen flowing into the rear of a conventional water-cooled oxy-cavity, it burns in the flame, causing it to become fuel-rich and to increase the amount of background emission. It also adds to the flow rate of gases transferred to the cavity and tends to entrain sodium hydroxide. This causes the cavity to emit the characteristic sodium emission at 589 nm which interferes spectrally with the NO–O emission. Such problems can be avoided by using a cavity in which nitrogen, the entrained products of reaction (NH_3 and H_2), and oxygen are mixed before ignition in a chamber behind the back of the cavity, as shown in Figure 3.5 (d).

The dimensions of the Devarda's alloy particles which have optimum reducing capacity were also studied. Belcher et al. (27) found that the most reproducible results and greatest emission intensities were obtained when these particles were less than 355 μm in diameter. To completely reduce nitrate in an aliquot (2 mL) of solution containing 1000 μg mL^{-1} nitrate, at least 30 mg of alloy was required. Larger amounts either did not increase the NO–O emission intensity, or they caused excessive amounts of hydrogen to be formed. When the latter occurred, the flame burned outside the MECA cavity and the signal intensity was decreased.

Different nitrate salts gave the same emission intensity within the concentration range studied for the same quantity of analyte. Measurement of peak areas rather than peak heights gave most consistent results, ten measurements of 100 μg mL^{-1} of nitrogen giving a relative standard deviation of 3.5%. The limit of detection was 1 μg mL^{-1} of nitrogen, while the calibration graph consisted of two distinct linear regions, one from 10 to 400 μg mL^{-1} and a second from 500 to 2500 μg mL^{-1}.

No interferences were observed for a nine-fold weight excess of Cu(II), Fe(II), Cd(II), Hg(II), Mn(II), Zn(II), Ni(II), Pb(II), Cr(III), Ag(I), Co(II), Cl$^-$, SO$_4^{2-}$,

CH_3COO^-, PO_4^{3-} or urea in the determination of 100 $\mu g\ mL^{-1}$ nitrogen as nitrate. Species such as nitrite, cyanide, metal-cyanide complexes and thiocyanate which also react with Devarda's alloy to form ammonia gave a positive interference. Arsenic and antimony also interfered by forming volatile hydrides with spectra overlapping that of NO–O (26).

A further extension of this study was undertaken by Al-Zamil and Townshend (36) who determined nitrite as well as nitrate. A potassium iodide reduction was used to convert nitrite to nitrogen monoxide, which was then transferred to the cavity for measurement. For nitrate determinations, one of two reduction methods was used. In the first, the nitrate was reduced initially to nitrite using zinc metal in alkaline solution as the reductant and Mn(IV) as the catalyst (37). The nitrogen monoxide formed by iodide reduction of this nitrite was then determined. In the second method, nitrate was reduced to nitrogen monoxide in one stage with zinc metal in acidic solution. The nitrogen monoxide was again quantified to determine the nitrate concentration.

In iodide reduction determinations, the system was purged with nitrogen gas to prevent any oxidation of nitrogen monoxide. The reaction vessel, containing 2 mL of 1 mol L^{-1} KI in 1 mol L^{-1} HCl, was maintained at either 96–100°C (NO_2^-) or 81–87°C (NO_3^-) and an aliquot of analyte (2 mL) added. As nitrogen monoxide was formed, it was swept into a trap pre-cooled to -196°C, located between the reaction vessel and the cavity. After 1 min the trap was taken out of the liquid nitrogen and re-immersed in a water bath at 80°C. Nitrogen monoxide was then transferred to the pre-heated stainless-steel oxy-cavity where the intensity of the emission at 526 nm was measured.

Blank measurements on nitrite and nitrate solutions injected directly into the cavity gave no emission at 526 nm, neither did nitrogen gas alone. It was concluded that the broad continuum centred at 526 nm was probably a composite of the NO–O spectrum reported earlier in NH_3 determinations (27) and that of NH_2.

In determinations of nitrite, the calibration graph was linear up to 300 $\mu g\ L^{-1}$ nitrogen. The detection limit was 0.5 $\mu g\ mL^{-1}$ while the relative standard deviation of seven replicates for 2 mL, each containing 100 $\mu g\ mL^{-1}$ nitrogen was 3%. No interferences were noted for five-fold weight excesses of Co(II), Cd(II), Hg(II), Pb(II), Ni(II), Zn(II), SCN^-, SO_4^{2-}, Br^-, PO_4^{3-}, HPO_3^{2-}, NO_3^-, $[Fe(CN)_6]^{3-}$ and $[Fe(CN)_6]^{4-}$. Highly oxidizing species such as Cr(III), Fe(III), BrO_3^-, MnO_4^-, CrO_4^{2-} and Ce(IV) depressed the NO–O emission to varying degrees because they converted the NO_2^- to NO_3^-.

In nitrate determinations using the iodide reduction method, solutions were maintained alkaline using sodium hydroxide instead of ammonia to prevent any possibility of spectral interference from the latter. Although conversion of nitrate to nitrite was 84–90%, consistent with values reported by Chow and Johnstone (37), the subsequent MECA determination of the nitrogen present proved to be somewhat irreproducible. The relative standard deviation of

Table 6.6. Reduction of 0.4 mg N as Nitrite or Nitrate by Some Granular Metals in 3.5 mol L^{-1} at 85°C as Measured by MECA

Metal (0.5 g)	Intensity (mV)		Metal (0.5 g)	Intensity (mV)	
	From Nitrite	From Nitrate		From Nitrite	From Nitrate
–	19.4	0.0	Zinc	45.0	22.8
Cadmium	36.0	3.2	Iron	40.5	17.0
Copper	45.0	2.8	Tin	28.5	0.0

five replicates of a solution containing 2 mL of 100 μg mL^{-1} nitrogen was 10%, a value much greater than that reported for the iodide reduction procedure for nitrite alone, or for the Devarda's alloy method of nitrite determination (27).

A more satisfactory procedure for nitrate determination involved the use of a reducing metal. Several such metals were investigated. A small amount of each (0.5 g) was placed in the reaction vessel, 3.5 mol L^{-1} HCl added and the nitrogen transfer apparatus reconnected. After de-aeration with nitrogen, and heating the reaction vessel to 84°C in a water bath, an aliquot of sample (2 mL) containing either nitrite or nitrate was added to the reductant and the nitrogen monoxide transferred to the cold trap. When all the analyte had reacted, the cold trap was warmed and the nitrogen monoxide was swept into the cavity where its emission intensity was measured.

Zinc and copper proved to be the most efficient reductants (Table 6.6). The former was therefore investigated more thoroughly for use in nitrite and nitrate MECA determinations. Using zinc as reductant, both nitrite and nitrate gave linear calibration graphs in the range 0–200 μg L^{-1} nitrogen with relative standard deviations for 2 mL of 100 μg L^{-1} nitrogen of 5%. The detection limits were 1 and 2 μg L^{-1} nitrogen for nitrite and nitrate, respectively. Negative interferences were observed where oxidizing agents (BrO$_3^-$, CrO$_4^{2-}$, MnO$_4^-$ and Ce(IV)) were present, but other species had no effect. These results are consistent with observations made with the iodide reduction method.

When nitrate was added to HCl in the reaction vessel without any metal being present, a significant but much decreased MECA signal was recorded. This lesser response was attributed to the presence of nitrogen dioxide as well as nitrogen monoxide in the cavity. The nitrogen dioxide either emits in the MECA cavity at a wavelength far removed from the NO–O emission, or it is incompletely broken down at the cavity temperature, or it is not transferred quantitatively to the cavity through the gas-phase delivery system.

Experiments were also conducted to determine nitrite and nitrate when present together. The concentration of nitrogen monoxide from both ions together was first measured by the zinc reduction method. A further sample of the same mixture was then treated with 1 mol L^{-1} KI in 1 mol L^{-1} HCl so that the nitrite

Table 6.7. Analysis of Nitrite and Nitrate Mixtures by MECA

Experiment No.	Expected (μg N mL^{-1})		Found (μg N mL^{-1})	
	NO$_2^-$	NO$_3^-$	NO$_2^-$	NO$_3^-$
1	100	100	104	97
2	50	100	48	98
3	100	50	95	53

was destroyed. The nitrate was then reduced by zinc metal and the nitrate-nitrogen determined by MECA. Nitrite was determined by difference. Results (Table 6.7) indicated that the method would be satisfactory for determining nitrite and nitrate at trace levels in admixture.

Celik and Henden (28) also studied the determination of nitrite and nitrate by MECA with the goal of applying the procedure to the analysis of meats and drinking water. Their procedure followed closely that of Belcher et al. (27) using iodide to reduce nitrite to nitrogen monoxide. The detection limit obtained, however, was significantly lower at 0.1 μg mL^{-1} nitrogen. Nitrate was determined by the iodide reduction method after first being reduced to nitrite in a copperized cadmium column (38,39). The efficiency of conversion of nitrate to nitrite for samples containing 0.5–10 μg mL^{-1} nitrogen was estimated to be 95–100%. Potential spectral interference problems associated with using an ammonia-based buffer were avoided by using a borate–boric acid buffer. This was only possible because the pH of the reagent solution could be varied between 5.3 and 9.6 without adversely affecting the reduction process. The calibration graphs and limits of detection for both nitrite and nitrate were similar.

As well as studying the same interferences discussed by Belcher et al. (27), Celik and Henden (28) considered the effect of possible interference from carbonate or sulphide. All observations made by Belcher et al. (27) were confirmed, but in addition Celik and Henden showed that carbonate and sulphide interfered through the production of carbon dioxide and hydrogen sulphide in the acidic environment of the reaction vessel. Both interferences were eliminated by placing a trap containing lead acetate, potassium hydroxide pellets and barium hydroxide powder between the reaction vessel and the cavity.

For meat analyses, samples were prepared according to standard procedures (40). Meat in soudjiks, salami and sausages was first minced, then a 10 g portion was extracted with hot water at 75–80°C. Proteins in the extracts were separated by precipitation with potassium hexacyanoferrate(II) and zinc acetate. Nitrite was determined directly in the extract using the specified reduction procedure combined with MECA. Recovery studies were made by adding known amounts of sodium nitrite to 10 g of the minced meat, and subjecting the

Table 6.8 Determination of Nitrite in Meat Products by MECA

	Amount of Nitrite (mg) per 10 g of Sample		
Sample	Found	Added	Recovered
Soudjik I[1]	0.085	0.33	0.31
Soudjik II[1]	0.21	4.11	4.03
Soudjik II[2]	0.21	4.11	3.90
Soudjik III[1]	N.D.	2.05	2.14
Soudjik III[2]	N.D.	0.82	0.79
Salami[1]	N.D.	2.05	1.95
Salami[2]	N.D.	2.05	1.99
Salami II[1]	N.D.	0.82	0.81
Salami III[1]	N.D.	0.82	0.81
Sausage I[1]	0.20	0.82	0.83
Sausage II[1]	0.12	2.05	1.94
Sausage III[1]	0.12	2.05	1.88

[1] Proteins precipitated before the measurements were made.
[2] Measurements made without separating the proteins. N.D. Not detectable.

mixture to the same analytical procedure. Results indicated that the procedure was quantitative for samples containing 10–60 mg kg^{-1} in nitrate. Further determinations on nitrite levels were also made without protein separation. These indicated that this particular separation was unnecessary for the MECA determination of nitrite-nitrogen in meats (Table 6.8).

Nitrate concentrations were determined by first reducing to nitrite any nitrate present in the extract from a 10 g meat sample, using a copperized cadmium column, determining the total nitrite-nitrogen by MECA, and then calculating the nitrate-nitrogen by the difference between this value and that obtained as nitrite-nitrogen from a separate 10 g meat sample. In this instance it was necessary to remove proteins from the extract before it was passed through the reductor column otherwise the column became blocked.

Table 6.9 Determination of Nitrate–nitrogen in Drinking Water by MECA. All Values in mg L^{-1}

Sample	MECA Technique	Spectrophotometric Method
1	1.40	1.30
2	1.52	1.48
3	6.80	6.80
4	5.20	5.20
5	1.72	1.68

Using the same general procedures, water samples were analysed for their nitrite and nitrate levels. No nitrite was detected but nitrate levels were discernible. These were compared to those obtained separately, using a spectrophotometric method of determination, and found to be similar (Table 6.9).

6.3 PHOSPHORUS COMPOUNDS

6.3.1 General

Phosphorus-containing compounds give the following gas phase emissions when they are excited:

(a) a line spectrum which occurs in the far ultraviolet,

(b) P–O bands which exist in the ultraviolet between 229.5 and 259.6 nm,

(c) a broad continuum from 490–645 nm with a maximum around 590 nm,

(d) the HPO band spectrum having a maximum at 527 nm with less intense maxima at 512 and 562 nm.

Of these, the most useful for routine analytical applications is the green HPO emission. It is relatively intense and because it occurs in the visible region of the spectrum, requires less demanding precautions for its observation. All volatile phosphorus-containing compounds as well as the vaporized element give this emission, which serves as the basis of the flame photometric detector for phosphorus in gas chromatography (41,42). It therefore provides excellent sensitivity for phosphorus determinations.

The main advantage of analytical methods based on emission spectroscopy rather than absorption spectroscopy for phosphorus-containing compounds is that the analyte does not first have to be converted to orthophosphate before it is determined. In the case of MECA, a bonus is that small sample volumes ($< 10 \mu L$) can be analysed.

Initially, the majority of MECA studies concentrated on organophosphorus compounds because of the relative volatility of the compounds and the ease with which they gave the HPO emission. More recently, studies have been undertaken to investigate the use of MECA in inorganic phosphorus compound analysis.

6.3.2 Inorganic Phosphorus Determinations

Early MECA studies on the determination of inorganic phosphorus compounds (43,44,45) using a stainless-steel cavity indicated that the majority of inorganic phosphates gave either a weak emission or no response. Phosphoric acid was the exception, giving a visible green emission forming two separate MECA peaks under relatively cool flame conditions but only one peak as the flame temperature increased. These emissions were attributed to and later confirmed as HPO

(8,43) and were thought to arise from vaporization of the orthophosphoric acid (first peak) and a polyphosphoric acid (second peak) formed within the cavity as it was heated. When a sample of orthophosphoric acid in the MECA cavity was pre-heated for 40 s, only the peak with the larger t_m value occurred (45). This appears to confirm that orthophosphoric acid is converted to another emitting species as it is heated up in the cavity, both of which give rise to HPO emissions.

Most metal phosphates are refractory compounds. At the temperatures necessary for their thermal breakdown ($>1000°C$), stainless-steel cavities incandesce and thus make impossible any sensitive observation of HPO emission. When experiments are undertaken at lower temperatures, the response is ill-defined and phosphate is incompletely evaporated from the cavity surface. Unless the residue is then washed out thoroughly, it builds up, retards volatilization of analyte in subsequent determinations, and makes their observation difficult.

Belcher et al. (44) partially solved this problem by replacing the stainless-steel cavity with one having a silica insert based on the model proposed by Syty (46). This was also found to be suitable for phosphoric acid studies.

Anionic interference studies showed that chloride could be tolerated in 10-fold weight excess while nitrate and sulphate when added in equal amounts to orthophosphoric acid had no effect. The HPO emission intensity was decreased by 50% when equal weights of borate and perchlorate were added to the analyte, while a 10-fold weight excess of these species reduced the emissions by 80% and 85%, respectively (45).

Sulphate proved interesting because the HPO peak sharpened as the amount of sulphate added was increased. With an extremely high level of sulphate (100-fold weight excess), two emission peaks were observed, one having a t_m associated with HPO, but the other clearly occurring earlier. This was attributed to S_2 whose spectrum, at the very high level of sulphate present, extended into the spectral region being investigated. The cavity produced an initial blue luminescence which appeared to confirm this assignment. In terms of cationic interferences, alkali metal and alkaline earth elements completely suppressed any HPO emission. This is likely due to refractory metal phosphate formation in the cavity through metathesis, for example:

$$3MCl_2 + PO_4^{3-} \longrightarrow M_3(PO_4)_2 + 6Cl^-$$

Ammonium ions had no effect on the emission intensity of HPO, even when present in 10-fold weight excess.

Attempts to decrease the level of interference from metal ions were partially successful when sulphuric or perchloric acid was added to the analyte. Sulphuric acid proved to be the better releasing agent.

Ammonium and sodium phosphates, sodium phosphite, sodium hypophosphite, barium hypophosphite, tetrasodium pyrophosphate and sodium polyphosphate, when prepared in 0.3 mol L^{-1} sulphuric acid, gave an HPO emission with the same t_m value as that of phosphoric acid. It has been suggested (45) that this observation is based on the fact that these phosphates are all hydrolysed to the same product, namely orthophosphoric acid, under acidic conditions. In this instance, calibration graphs for ammonium phosphate, phosphoric acid, sodium tripolyphosphate and calcium phosphate were all linear in the range 5–50 µg mL^{-1} phosphorus. For calcium phosphate, the limit of detection was 5 µg mL^{-1} phosphorus with a coefficient of variation of 2.8% (45).

These results have since been questioned by Knowles *et al.* (47) who found that the concentration of sulphuric acid used by Osibanjo to release HPO emissions from metal phosphates (0.3 mol L^{-1}) (45) also gave rise to an interfering sulphur emission at 528 nm. The dilemma was resolved by converting each phosphorus-containing species to the corresponding ammonium or acid form with an ion-exchange resin, and then analysing each in a carbon cavity, using the HPO emission.

Two further interesting points emerged from this study. The first is that the t_m values for phosphorus emissions from a variety of different phosphorus-containing molecules treated in this fashion are very similar. This means that the t_m value alone cannot be used to identify the analyte. Secondly, emission intensities for a variety of phosphorus-containing compounds at the same concentration were similar. Linear calibration graphs for the compounds treated by ion-exchange were obtained in the range 0–50 µg mL^{-1}. Thus, although the MECA procedure cannot be used to determine different inorganic phosphorus-containing compounds, it would be useful for the determination of overall inorganic phosphorus levels.

The interference of arsenic in this method was also studied. In contrast to the spectrophotometric determination of phosphorus where arsenic gives a positive interference, arsenic suppresses the MECA emission and is therefore a negative interferent. However, this effect is unimportant when the arsenic level in solution ≤1%.

Studies on metal phosphates and phosphoric acid by Krnac *et al.* (48), using a platinum insert within a stainless-steel cavity and argon rather than nitrogen as the flame diluent, have confirmed these findings. In addition, they noted that H_2O_2 added to the sample increased the emission intensity whereas air entrained in the flame gases had no effect. Ammonium ions in the absence of sulphuric acid had no effect on the signal intensity, but ammonium salts of phosphoric acid provided a signal enhancement of 20% over that obtained from equimolar concentrations of phosphoric acid alone. They suggested that the enhancement of HPO emissions from ammonium salts was probably due to spectral overlap of the NO–O continuum with the HPO spectrum. Calibration graphs were non-linear in this system and the detection limit was 0.1 µg phosphorus. The reproducibility was determined to be 3.8%. After 400 or so determinations with the same cavity, it started to lose its brightness and

Table 6.10. The Effect of Various Cations on the HPO Emission Intensity from 100 mg L^{-1} P before and after Treatment with a Cation Exchange Resin

Interferent[1]	Relative Emission intensity[2]			
	Interferent Alone		Interferent + 100 mg L^{-1} P	
	Before	After	Before	After
0.01 mol L^{-1} Na^+	4.5	0	4.5	77.3
0.1 mol L^{-1} Na^+	50.0	22.7	50.0	22.7
0.01 mol L^{-1} K^+	18.2	0	22.7	91.0
0.1 mol L^{-1} K^+	109	45.4	109	50
Ba^{2+}	0	0	78.8	100
Ca^{2+}	0	0	34.6	100
Cu^{2+}	63.1	0	105	100
Cr^{3+}	0	0	10.5	100
Co^{2+}	0	0	89.5	100
Mn^{2+}	5.3	0	63.1	100
Al^{3+}	0	0	0	100
Fe^{3+}	2.6	0	68.4	100

[1] All added as chlorides. Apart from alkali metals, all metals were added at 100 mg L^{-1} concentrations.
[2] The emission intensity from the pure solution of 100 mg L^{-1} P (1.30 mV) arbitrarily taken as 100.

emission intensities decreased. The emission intensity was also less when dust from the laboratory was allowed to contaminate the cavity surface. Undoubtedly, the nature of the cavity surface plays an important role in determining the emission intensity from any species being analysed by the MECA technique.

Recently, Nakajima et al. (49) have made a more thorough investigation of how cavity surface affects S_2 emissions. Using the techniques of X-ray and electron probe microanalysis, they have shown that the cavity surface giving the highest emission intensity is one that is rough. According to these investigators, this type of surface allows good dispersion of analyte, excellent volatilization of sample, and suppression of peak splitting.

Another example of how ion-exchange resins can be used to good effect in phosphorus MECA determinations has been provided by Calokerinos and Hadjiioannou (50). They utilized a batch process in which the resin was first added to the appropriate analyte solution, stirred for 15 min, and then filtered. The filtrate was aspirated into a hydrogen–nitrogen flame, as reported in an earlier study (51). Emission intensity suppression from Ba(II), Ca(II), Cu(II), Cr(III), Co(II), Mn(II), Al(III) and Fe(III) was completely eliminated by this procedure (Table 6.10). Interferences from sodium and potassium ions, however, were only partially removed.

Table 6.11. Determination of Phosphate in Phosphate-bearing Rocks[1]

P_2O_5 (%) present	17.35	22.80	25.19	26.69	32.80
Found	17.50	22.97	24.67	27.32	31.78
Relative error (%)	+0.9	+0.7	−2.1	+2.4	−3.1

[1] T. Smith Analysed Standards.

Although hydrochloric acid did not affect the emission from the filtrate, sulphuric acid (as noted previously (47)) and nitric acid proved to be serious interferences. Nitric acid decreased the HPO emission intensity, probably because NO radicals present in the flame gas atmosphere catalysed the recombination of atomic hydrogen and hydroxyl radicals to water vapour with a subsequent decrease in flame radical concentration (52). Direct comparison of emission intensities from phosphorus-containing compounds prepared in nitric acid and in water, therefore, are not possible. This problem can be avoided, however, if a standard addition procedure is used for samples prepared in nitric acid.

More practical applications of MECA to the determination of inorganic phosphorus-containing compounds have included the analysis of lecithin (48), rocks (50) and detergents (53,54). Krnac et al. (48) determined phosphorus in technical lecithin and showed that a procedure based on MECA gave results comparable to those obtained using more conventional spectrophotometric means. The quantities of lecithin used in this study were such that they gave HPO emission intensities falling on the linear portion of the orthophosphoric acid calibration graph (0–60 $\mu g\ mL^{-1}$ P). Results of five replicate determinations on each sample gave a relative error of −1.8%.

In their study of rock analyses by MECA, Calokerinos and Hadjiioannou (50) found good agreement between the values for phosphorus obtained by this technique and those certified as correct by the supplier (Table 6.11). These authors also reported excellent recoveries (97–103%) from phosphate added to liquid detergents.

Osibanjo et al. (53) used the HPO emission to determine total phosphate in several detergents. These were first ashed, then dissolved in sulphuric acid. Aliquots of the resulting solutions were injected into a carbon cavity which was heated in a hydrogen–nitrogen–air flame and the HPO emission intensity measured. This was compared to the emissions from calibration standards prepared from sodium dihydrogen phosphate to obtain the concentration of phosphorus present in the detergent. Results indicated that the MECA procedure compared favourably with the molybdenum blue and quinoline molybdate spectrophotometric methods (53,54) (Table 6.12).

The question of reproducibility was addressed by El-Hag and Townshend (55) who developed an automated MECA system. The system, itself, is more fully described in Chapter 2. El-Hag and Townshend, working with a number

Table 6.12 Determination of Phosphorus in Detergents

Detergent Samples	% Phosphorus		
	MECA	Molybdenum Blue	Quinoline Molybdate
Daz	7.5, 7.4	7.8, 7.6	7.3, 7.9
Drive	9.6, 9.6	8.3, 8.4	9.4, 9.8
Omo	8.3, 8.3	7.2, 7.3	8.2, 8.1

of sodium compounds of phosphorus anions, first eliminated any possible interference from the cation by treating the analyte with an ion-exchange resin (Dowex 50W-X8 cation-exchanger). The resulting sodium-free solution was placed in a vial from which aliquots (5 μL) were injected automatically into a carbon cavity, held in a 'loading' position. The cavity was then moved into the 'measuring' position in the flame, where it was held for 10 s while the emission intensity at 528 nm was measured. Subsequently, the cavity was removed from the flame and allowed to cool for 90 s before the cycle was repeated. The complete operation for injection of analyte and placement of cavity was computer controlled.

The calibration graphs for ortho-, pyro- and tripolyphosphate were almost identical. Those for tri- and tetrametaphosphate were also similar. It was concluded that because of these similarities and since the t_m values for each of these compounds was the same, the major species generating an HPO emission was common to all of them and was most likely orthophosphate (45). For phosphite and hypophosphite, which give rise to more volatile products than the phosphates, increased sensitivity and a lower t_m value were observed. This increase in sensitivity at lower cavity temperatures is consistent with results observed for sulphur compounds (see Chapter 4) (1).

The major advantage of the automated method over the manual one was a major increase in reproducibility. For phosphates, the relative standard deviation for 10 replicate determinations of 250 ng was 0.9%, while for phosphites it was 2.6%. This compared with 4.5% and 10–20%, respectively, using manual sample introduction. Limits of detection were 2.5 ng of phosphorus for phosphite and 7.5 ng of phosphorus for phosphate in a 5 μL sample.

The technique was applied to the determination of phosphorus in fertilizers. A sample of fertilizer was first digested for 30 min in water. After filtering this suspension, the filtrate was diluted to volume and a portion treated with the cation-exchange resin for 5 min in a batch process. Aliquots of the supernatant layer (5 μL) were then analysed in the automated MECA instrument. The amount of phosphorus present in the fertilizer sample (Standard Sample No. 1 for Fertilizer Analysis, Fertilizer Manufacturer's Association) was determined by MECA to be 4.8%, the same as that on the authenticity certificate. The relative standard deviation was 3.6% for ten replicates.

Although the fertilizer contained sulphur as well as phosphorus and the blue S_2 emission was observed in addition to the HPO emission, its effect could be eliminated by changing the position of the cavity in the flame from the centre, where it was cooler, to the outer edge where the temperature was higher. This procedure appeared to negate the effect of the Salet phenomenon for S_2 (56), while having a much more limited effect on the HPO emission.

6.3.3 Organic Phosphorus Determinations

Many organophosphorus compounds are relatively volatile and readily decomposed by heat. When heated in a cool flame, they often generate the HPO emission. These features have made them attractive as a group to study by MECA.

Osibanjo (45), using an ethanolic solution of tri-n-butyl phosphate, found that a fuel-rich hydrogen–air flame, diluted with nitrogen, provided the best MECA environment for stimulating HPO emissions from these compounds. He also compared aluminium and stainless-steel as cavity materials and found the former to be the more satisfactory since it conducted heat away from the cavity faster.

Solvent evaporation time was also found to be important. Best results were obtained when the ethanol was allowed to evaporate for 10 s before the MECA cavity was heated in the flame. When the ethanolic solution was heated in the cavity without prior solvent evaporation, the HPO emission intensity was decreased. He found that evaporating the solvent for 10 s before making a measurement of HPO emission intensity was the procedure that gave the greatest response for most organophosphorus compounds. However, comparatively volatile materials such as trimethyl phosphite would evaporate within 10 s and the evaporation stage had to be shortened accordingly.

Most compounds studied gave a single emission response when measured at 528 nm whether or not the solvent was pre-evaporated. Triphenyl phosphite was different, however, in that solvent evaporation time was important in determining the shape of the emission profile. When a solution of triphenyl phosphite was heated without prior solvent evaporation, the MECA signal consisted of one peak. By allowing the solvent to evaporate first for 10 s, however, the emission response split into two, the first component having a t_m from 1.5 to 3.0 s depending on how long the solvent had been left to evaporate, and the second having a constant t_m of 5.0 s. It was concluded that this behaviour was due to the oxidation of triphenyl phosphite to triphenyl phosphate within the cavity. The first peak was assigned to the parent compound while the second was attributed to the oxidation product.

Besides considering cavity material and solvent evaporation time, Osibanjo also concluded that the magnitude and t_m of the emission response was dependent on the solvent used and the organophosphorus compound being investigated. Better sensitivities were obtained using ethanol and benzene rather than methyl isobutyl ketone as solvent. The order of sensitivity in both stainless-steel and aluminium cavities was phosphines $\geqslant$ phosphine oxides > phosphites > phosphates, indicating a decrease in emission response as the oxygen content of the molecule increased.

These observations were explained on the basis of relative bond strengths. The hydrogen-based flame is comparatively cool, so has difficulty in breaking phosphorus–oxygen bonds (P=O: 585 kJ mol^{-1}; P–O: 397 kJ mol^{-1}). Phosphorus–carbon bonds, on the other hand (P–C: 263 kJ mol^{-1}), dissociate more easily.

Linear calibration graphs were obtained for triphenyl phosphine, tri-n-butyl phosphate, triphenyl phosphine oxide, trimethyl phosphite and triethyl phosphate, although the range was no greater than 0–40 μg mL^{-1}, and, in the case of the more volatile compounds, was considerably less. Detection limits varied from 0.1 to 2.0 μg mL^{-1} with reproducibility for 10 replicates being 2–3% for samples containing 2–3 ng of analyte (45).

An evaluation of how water cooling of the cavity affects the MECA emission response was made by Belcher et al. (57). Using a water-cooled aluminium cavity, they investigated changes in HPO emission intensity and t_m values of 12 organophosphorus compounds as the water flow rate was increased. Figures 6.3 and 6.4 present data for these compounds.

Trimethyl phosphate, triethyl phosphate, triphenyl phosphite and trimethyl phosphite, compounds that have comparatively low boiling points, showed enhanced emission intensity (2–6 times) as the water flow rate increased and cavity temperature decreased. Calibration graphs remained linear. Other organophosphorus compounds, such as tritolyl phosphate, triphenyl phosphate, di-(2-ethylhexyl) phosphate, tri-n-butyl phosphate, triphenyl phosphine and triphenyl phosphine oxide showed a steadily decreasing emission intensity, as the rate of water cooling increased. These compounds have such a low volatility at room temperature that they were unable to provide any HPO emission when the cavity was well cooled in this fashion.

The effect of cavity temperature on individual t_m values is less clear. As with sulphur-containing compounds, however, so long as the t_m value can unambiguously be assigned, it can be used to identify the compound giving rise to the HPO emission. Water-cooling of the cavity generally assists this process by either enhancing the emission through the Salet phenomenon (56) for more volatile species, or decreasing the intensity to zero for less volatile compounds. This can be seen in Figure 6.5 where the effect of cavity cooling on the emission responses of trimethyl phosphate, triphenyl phosphate and di-(2-ethylhexyl) phosphate is shown.

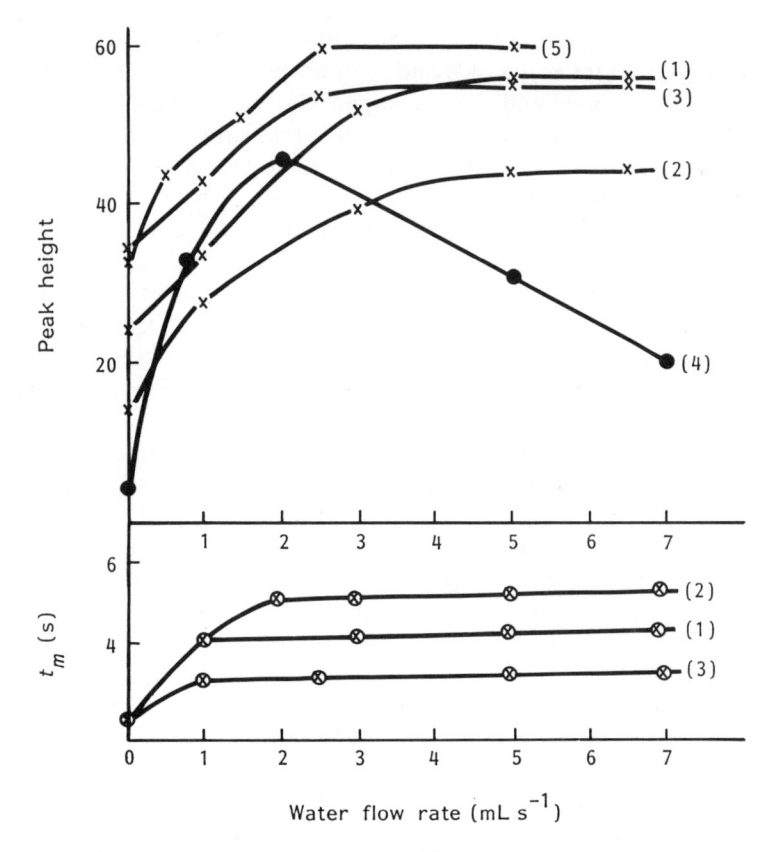

Figure 6.3. Effect of water-cooling on the MECA response characteristics of: (1) trimethyl phosphate (100 μg mL^{-1} P); (2) trimethyl phosphite (20 μg mL^{-1} P); (3) triethyl phosphate (25 μg mL^{-1} P); (4) diethyl phosphite (12 μg mL^{-1} P); (5) triphenyl phosphate (20 μg mL^{-1} P).

In separate investigations, Wang and Deng (58) and Krnac and Heftjmanek (59) confirmed the findings of Belcher *et al.* (57). The latter increased the number of organophosphorus compounds and solvents studied and changed the experimental technique by using a platinum cavity insert and argon instead of nitrogen as one of the flame gases (48). Besides observing that the strength of phosphorus bonds to other atoms in the molecule under investigation were important in determining the MECA response of organophosphorus compounds, Krnac and Hejtmanek found that functional groups also had an effect. For instance, the weak phosphorus emission intensity from chlorotriphenylphosphine compared to that from triphenylphosphine was attributed to the relatively weak P–Cl bond (227.7 kJ mol^{-1}) which hydrolysed in the

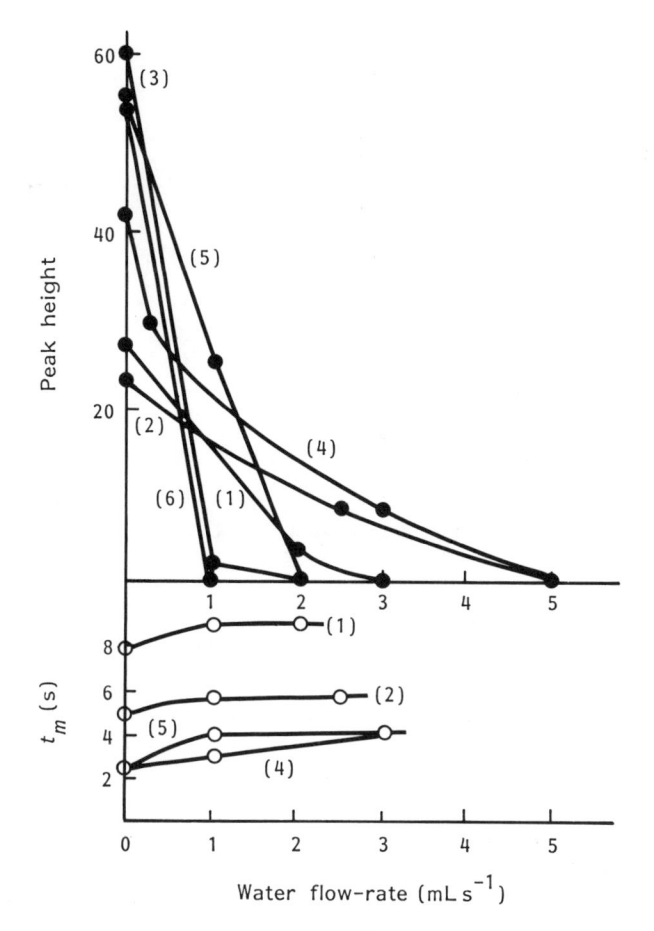

Figure 6.4. Effect of water-cooling on the MECA response characteristics of: (1) tritolyl phosphate (50 μg mL^{-1} P); (2) triphenyl phosphate (50 μg mL^{-1} P); (3) di-(2-ethylhexyl) phosphate (100 μg mL^{-1} P); (4) tri-n-butyl phosphate (5 μg mL^{-1} P); (5) triphenylphosphine (5 μg mL^{-1} P); (6) triphenylphosphine oxide (40 μg mL^{-1} P).

humid air of the laboratory to a more stable P–O linkage:

$$(C_6H_5)_2PCl + H_2O \longrightarrow (C_6H_5)_2POH + HCl$$

This investigation was also extended to a study of MECA responses from organophosphorus insecticides, namely Araforsfotion, containing 26% malathion (*O,O*-dimethyl-5-(1,2-dicarboethoxyethyl)dithiophosphate) and Metation E-50, which contains approximately 50% fenitrothion (*O,O*-dimethyl-*O*-(3-methyl-4-nitrophenyl)thiophosphate). Each insecticide was studied in aqueous and chloroform solutions.

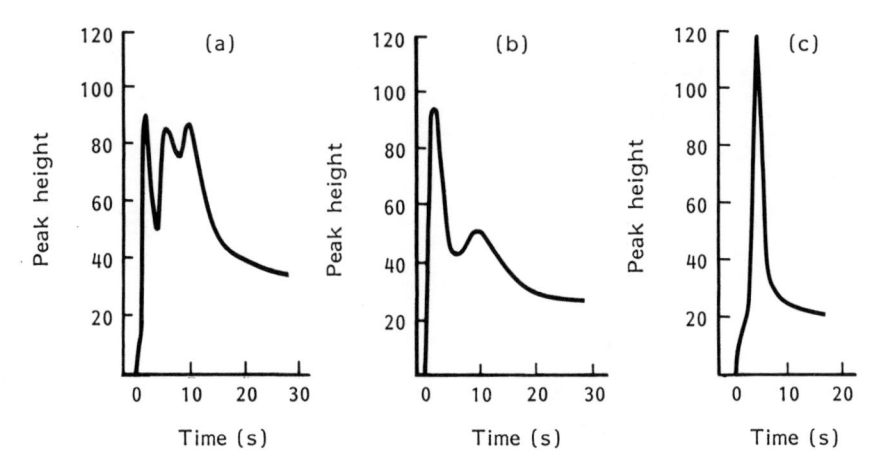

Figure 6.5. Effect of water cooling on the MECA response from a mixture of trimethyl phosphate (100 μg mL^{-1} P), triphenyl phosphate (50 μg mL^{-1} P) and di-(2-ethylhexyl) phosphate (50 μg mL^{-1} P); (a) no water flowing; (b) 1 mL s^{-1} water flow; (c) 10 mL s^{-1} water flow.

Little sulphur interference on the phosphorus MECA response was noted from either insecticide, contrasting sharply with the effect of much higher concentrations of sulphur on the phosphorus MECA emissions from phosphoric acid (48). A single well-defined peak (t_m = 4 s) was observed over the whole concentration range studied, with detection limits of 2 and 3 ng mL^{-1} phosphorus for Araforsfotion and Metation E-50, respectively. The choice of solvent had no effect on these values.

An interesting application of organophosphorus MECA has been reported by Belcher et al. (60) who used it as the basis of a gas chromatography (GC) detector. Subsequently, Cope and Townshend (61) applied phosphorus MECA detection to liquid chromatography (LC).

In the GC study (60), trimethyl phosphate and triethyl phosphate were separated on a PTFE column packed with 10% Silicone Gum Rubber E-301 on Porapak Q (50–80 mesh), with nitrogen as carrier gas. The column was connected by a 30 cm PTFE tube directly to the rear opening of either a water-cooled conventional or oxy-cavity constructed of Duralumin. This tube and the injection port to the cavity were maintained at 20 K above the column temperature by wrapping them with an electrically-heated tape. Measurements of emission intensity were made at 526 nm. Individual MECA responses of eluted analytes (Figure 6.6) showed complete separation on the column. The calibration graph for trimethyl phosphate (peak height) was linear within the range 0.05–1.3 μg of P which was the same as that for triethyl phosphate (peak area). The likely cause of curvature of the calibration graph for trimethyl phosphate above 1.3 μg of P is self-absorption. Coefficients of variation for the determination of 0.4 μg of P

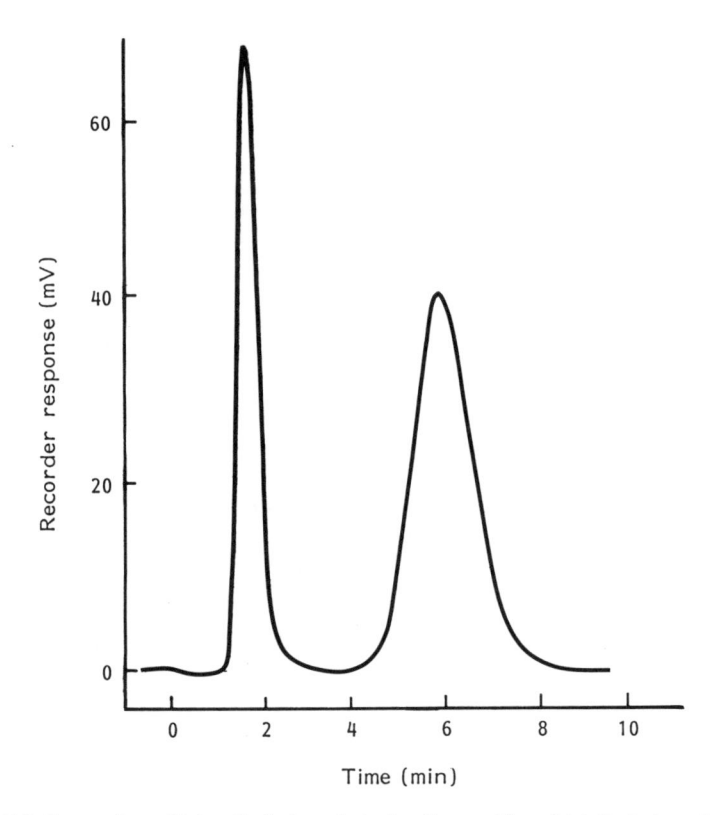

Figure 6.6. Separation of trimethyl phosphate (earlier peak) and triethyl phosphate, with MECA detection at 526 nm.

as trimethyl phosphate and triethyl phosphate (eight replicates) were 1.7 and 2.0%, respectively. The limit of detection for trimethyl phosphate was 2 ng of phosphorus. A search for spectral interference of S_2 (λ_{max} = 384 nm) and CH (λ_{max} = 431.5 nm) on the phosphorus emission showed that neither carbon nor sulphur interfered significantly. Co-elution of organic compounds containing carbon but not phosphorus had the effect of quenching the HPO emission (45).

In their LC–MECA study, Cope and Townshend (61,62) developed a method of sample analysis where the eluent from the LC column was analysed directly. As each droplet of eluent emerged it was deposited into a cavity drilled into a Duralumin disc, whereupon the disc rotated to allow the next droplet to be deposited into another cavity (Figure 3.12) (Chapter 3).

Using a nominal wavelength of 528 nm and a 3 nm spectral bandpass, experiments to determine the MECA sensitivity of these cavities were undertaken with aqueous solutions of triethyl phosphate. This gave a linear calibration graph

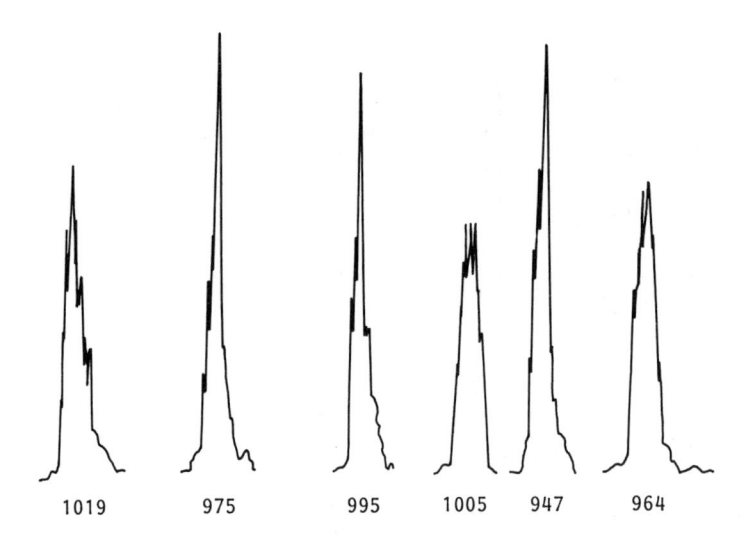

Figure 6.7. Liquid chromatogram from six identical injections of 8 μg of triethyl phosphate. Retention time 8.5 min. Lichrosorb 10 μm adsorption column (25 cm × 3.2 mm i.d.); mobile phase, methanol at 0.2 mL min^{-1}; 400 psi; water-bath temperature, 30°C; chart speed, 30 m min^{-1}. The numbers below the peaks denote their relative peak areas.

within the range 10–100 ng phosphorus with a 2σ detection limit of 5 ng. Peak areas rather than peak heights were used in LC-related measurements because they showed less variability.

When identical amounts of triethyl phosphate (8 μg) were passed through an adsorption column (Lichrosorb) using methanol as the mobile phase, the MECA responses from composite peak area measurements varied by no more than 3%. Individual peak heights from a given eluate varied by considerably more than this (Figure 6.7) because of the variability of splitting the eluting compound into individual drops.

When tributyl phosphate, tripropyl phosphate and triethyl phosphate were applied to the same column as a mixture, using a mobile phase of hexane:isopropanol (80:20), the MECA responses showed that the components had been resolved satisfactorily (Figure 6.8). However, one problem that required attention was the presence of a memory peak (m) caused by the incomplete evaporation of residual phosphate in the cavity. This proved not to be significant, though, because the 'memory' peak had a t_m sufficiently far removed from those of the analytes that it did not interfere.

Other phosphorus-containing compounds studied included a mixture of dibutyl methylphosphonate, diisopropyl methylphosphonate and diethyl methylphosphonate and one of diethyl dimethylpyrophosphonate, dipinacolyl dimethylpyrophosphonate and dibutyl methylpyrophosphonate, separated in

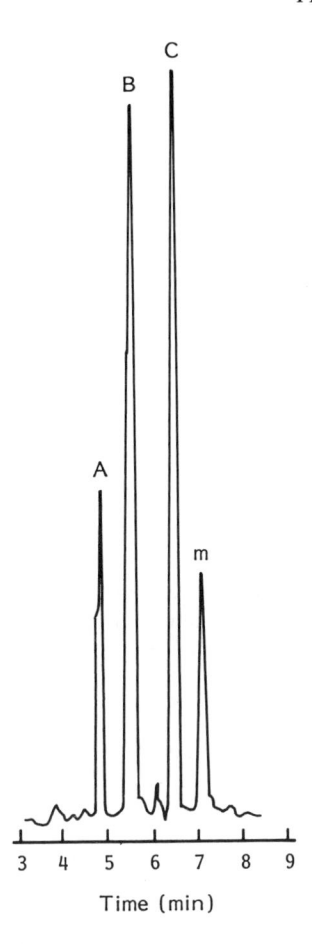

Figure 6.8. Phosphorus-selective liquid chromatogram from Partisil 5 μm adsorption columns (25 cm × 4.6 mm i.d.). Chromatogram of (A) tributyl phosphate (3.15 μg P), (B) tripropyl phosphate (3.13 μg P) and (C) triethyl phosphate (1.67 μg P); hexane: isopropanol (80:20) mobile phase at 0.9 mL min^{-1}; 430 psi; index time 2.2 s; water bath temperature 36°C. Peak marked (m) is a memory peak.

each case by using a mobile phase of hexane:ethanol (80:20). These were also clearly identified from the MECA responses of the eluates; again the t_m values were distinguishable from those of 'memory' peaks (Figures 6.9 and 6.10).

In the case of phosphonic acids, where LC separation was achieved by using reversed-phase rather than adsorption chromatography, the cavity-bearing disc had to be heated to 90°C to ensure greatest sensitivity. Water-cooling of the cavity was unnecessary. Figure 6.11 shows the HPO MECA response from a chromatographic separation of methylphosphonothionic and ethyl methylphosphonic acid using formic acid (0.1 mol L^{-1}; pH 2.55) as the mobile phase.

The technique for introducing analyte into the cavities at this elevated temperature had to be modified. At 90°C droplets of analyte did not wet the surface. The droplet was insulated from the cavity surface by a bubble of solvent vapour and therefore gave low MECA responses. A more satisfactory

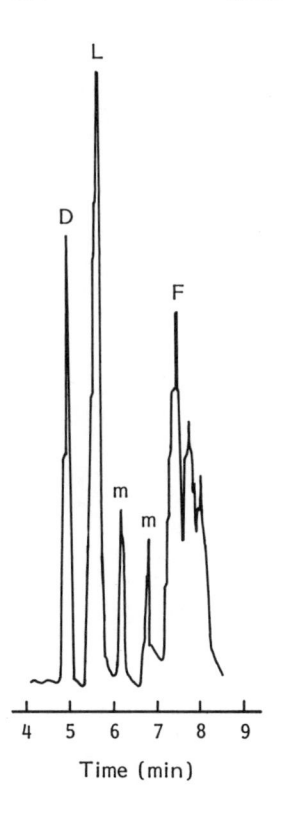

Figure 6.9. Phosphorus-selective liquid chromatogram from Partisil 5 μm adsorption columns (25 × 4.6 mm i.d.). Chromatogram of (D) dibutyl methylphosphonate (2.13 μg P), (E) diisopropyl methylphosphonate (2.8 μg P) and (F) diethyl methylphosphate (5.09 μg P); hexane:ethanol (80:20) mobile phase at 0.9 mL min^{-1}; 460 psi; index time 1.9 s; water bath temperature 40°C. Peaks marked (m) are memory effects.

procedure which caused the cavity surface to become wetted and which restored the emission intensity, was to increase the eluent flow rate so that eluent was sprayed rather than dripped from the LC column.

It is clear that the MECA technique is suitable for establishing the presence of organophosphorus compounds in eluents from GC and LC with a variety of mobile phases. So far, however, it has not been considered for more general use, probably because the more conventional photometric detectors are more sensitive and more reproducible, at least for GC.

6.4 CARBON COMPOUNDS

In their study concerning the determination of carbonyl compounds, Al-Abachi et al. (63) recorded the S_2 response from sulphite addition compounds formed between the appropriate aldehyde or ketone and sodium sulphite. This response comprised two peaks, the first being due to unreacted excess sulphite, and the second to the sulphite addition compound. By measuring the height of the second peak, carbonyl compounds such as formaldehyde (2–750 μg), acetalde-

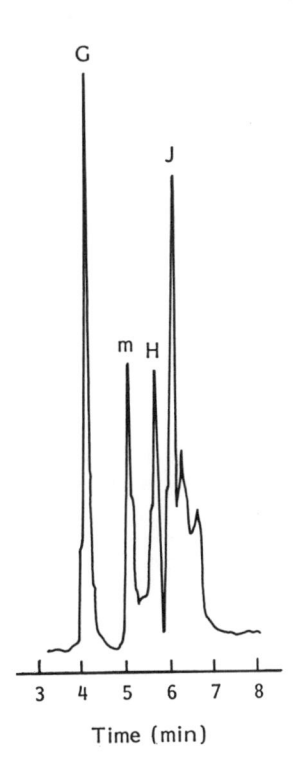

Figure 6.10. Phosphorus-selective liquid chromatogram from Partisil 5 μm adsorption columns (25 cm × 4.6 mm i.d.). Chromatogram of (G) diethyl dimethylphosphonate (2.74 μg P); (H) dipinacolyl dimethylphosphonate (3.6 μg P); and (J) dibutyl dimethylpyrophosphonate (1.86 μg P). Mobile phase as in Figure 6.9; index time 1.36 s; water bath temperature 37.5°C. Peaks marked (m) are due to memory effects.

hyde (0.05–1.0 mg) and acetone (0.4–3.5 mg) could be determined. Mixtures of formaldehyde and acetone gave resolved S_2 emission responses and were determined simultaneously after evaporation of excess sulphite from the cavity. Some aromatic aldehydes such as benzaldehyde, p-dimethylaminobenzaldehyde, p-chloroaminebenzaldehyde and p-nitrobenzaldehyde were also determined with detection limits in the range 1.0–1.7 ng per 5 μL aliquot (64).

In another investigation, Al-Ghabsha (64) determined various alcohols by first converting them to the appropriate carbonyl compound through periodate oxidation, and then quantifying them from the S_2 emission of a sulphite addition compound. Since periodate and/or iodate oxidizes some of the sulphite to sulphate, three peaks were recorded. In order of increasing t_m values, these were: unreacted sulphite; addition compound; and sulphate produced through oxidation. Some compounds gave more than three peaks because more than one oxidation product was formed. For instance, 2-amino-2-methyl-1-propanol and 1-phenyl-1,2-ethanediol produced acetone and formaldehyde, and benzaldehyde and formaldehyde, respectively (Figure 6.12).

Al-Zamil et al. (65) have determined cyanide indirectly through its reaction with the sulphite addition compound of formaldehyde. The increase in emission intensity from reduced sulphite (t_m = 6 s) and the decrease in intensity

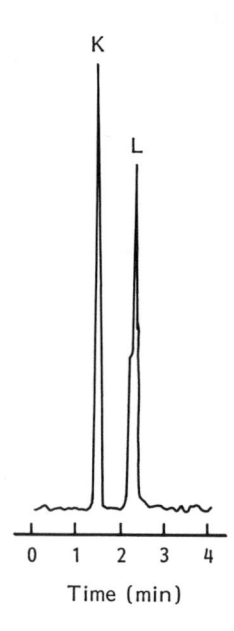

Figure 6.11. Phosphorus-sensitive liquid chromatogram of (K) methylphosphonothionic acid (10.8 μg P) and (L) ethyl methylphosphonic acid (15.7 μg P). Conditions: Partisil 10 μm reversed-phase ODS column (10 cm × 5 mm i.d.); 0.1 mol L^{-1} formic acid mobile phase at 2.0 mL min^{-1}; 1040 psi; no water cooling.

of the addition compound (t_m = 15 s) were both proportional to the amount of cyanide present, so either emission intensity could be used for cyanide determination in nanogram amounts.

Nitrite ions completely suppressed the sulphite emission and also reduced the height of the addition compound peak. This effect was attributed to the complete oxidation of sulphite and partial oxidation of the addition compound by the nitrite. Nitrate ions (240 ng) showed no significant effect on the determination of cyanide (100 ng). Carbonate and halide ions liberated sulphite from the addition compound and interfered. Addition of urea and nitric acid to the sample solution, prior to analysis, was suggested as a means of avoiding interferences from nitrite and carbonate, respectively.

Al-Zamil and Townshend (66) have determined ethanol via its oxidation to acetaldehyde by nicotinamide adenine dinucleotide (NAD$^+$) in the presence of yeast alcohol dehydrogenase (ADH). The aldehyde was then converted to the corresponding sulphite-addition compound by the addition of sodium sulphite, and the amount of ethanol determined indirectly by MECA from the sulphite-addition compound's S$_2$ emission intensity.

$$NAD^+ + C_2H_5OH \xrightarrow[\text{pH 7-8}]{\text{ADH}} NADH + CH_3CHO + H^+$$

In this investigation, standard solutions of ethanol within the range 0–240 ng per 5 μL, or an 'unknown' sample were mixed with aqueous solutions of NAD$^+$ and Tris buffer (pH 8.0) and held at 25°C for 10 min. Yeast ADH was added

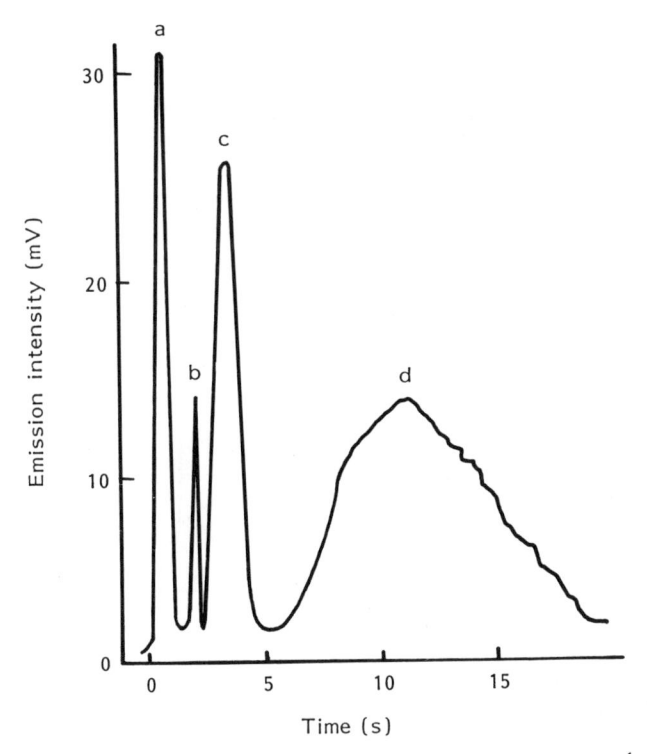

Figure 6.12. S_2 emission profiles from 5 µL solution containing 120 µg mL^{-1} of 2-amino-2-methyl-1-propanol treated with 0.0024 mol L^{-1} NaIO$_4$, then with 0.012 mol L^{-1} Na$_2$SO$_3$ and 0.06 mol L^{-1} H$_3$PO$_4$, showing responses from (a) excess sulphite, (b) acetone–sulphite addition compound, (c) formaldehyde–sulphite addition compound, and (d) sulphate.

and the reaction allowed to proceed for exactly 10 min. At the end of this period, the acetaldehyde produced was treated with excess sodium sulphite solution containing disodium-EDTA as stabilizing agent, and left at room temperature for 5 min. A small portion of phosphoric acid was added and the solution diluted to volume. Aliquots (5 µL) of this solution were injected into a carbon cavity, heated in a nitrogen-cooled hydrogen flame, and the intensity of the S_2 emission at 384 nm measured after evaporating any free sulphur dioxide with an air blower. The cavity was then heated in an air-entrained flame to burn off any carbon deposits from its inner surface, after which the cycle was repeated.

Carbon rather than stainless-steel cavities were used in this study because they were found to give more reproducible results, even though the resolution of peaks was less clear. Resolution of the emission responses from sulphur dioxide (t_m = 2 s; peak [a], Figure 6.13) and sulphite-addition compound

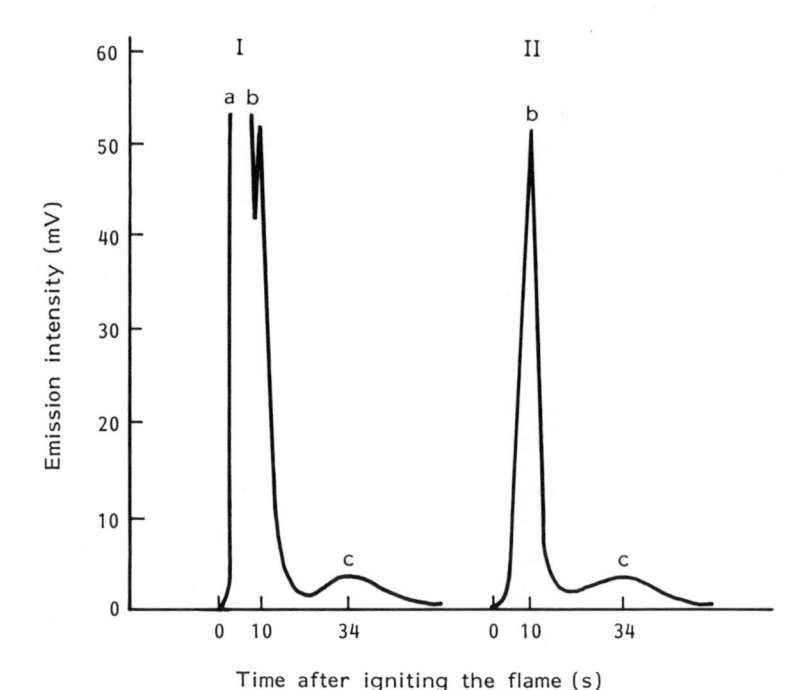

Figure 6.13. MECA response from 230 ng ethanol per 5 μL without (I) and with (II) evaporation of excess of SO_2. Peaks: (a) free SO_2 (t_m = 5 s); (b) sulphite addition compound (t_m = 10 s); (c) sulphate (t_m = 34 s) (carbon cavity).

(t_m = 8 s; peak [b], Figure 6.13) had no significant impact on the determination provided sulphur dioxide was pre-evaporated from the sample in the cavity. The remaining peak (t_m = 34 s) was ascribed to sulphate, formed by air oxidation of sulphite within the cavity.

Experiments to determine the length of time necessary for the enzyme to react completely showed that it had to be left in contact with the substrate for 6–8 min (Figure 6.14). When sulphite was added to the reaction mixture before it had undergone complete reaction, the S_2 emission intensity was decreased by 20%, probably because the enzyme action was inhibited by the sulphite. In all experiments with ethanol, the ADH and NAD^+ concentrations were relatively high to ensure a fast and complete reaction. The plot of emission intensity against ethanol concentration was linear over the range 0–240 ng per 5 μL injection. The detection limit was established at 10 ng of ethanol, and 'unknowns' containing 46, 92 and 184 ng ethanol per 5 μL of solution analysed at 47, 96 and 190 ng, respectively.

Attempts were made to broaden the scope of the study by using 1 mol L^{-1} methanol, n-propanol and isopropanol instead of ethanol. However, these

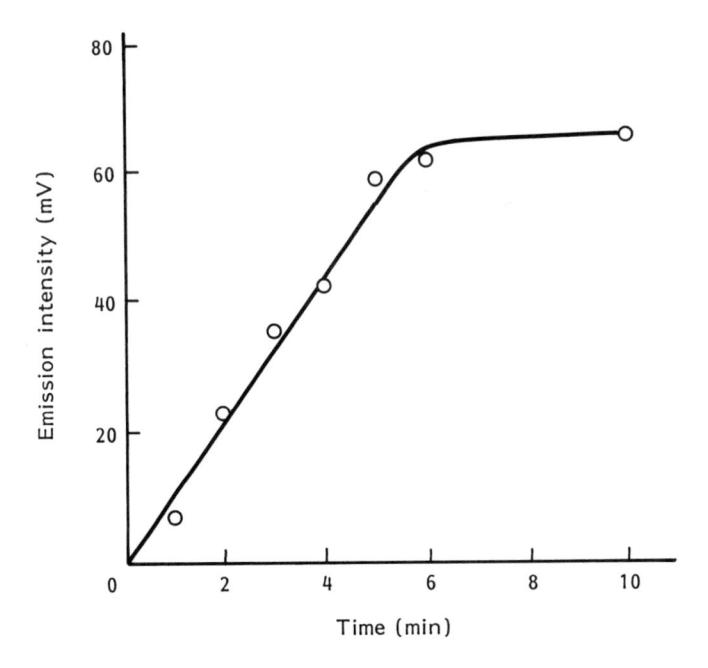

Figure 6.14. Effect of time on the reaction of ethanol (230 ng per 5 μL) under recommended conditions.

alcohols gave an insufficiently strong emission response for this method to be considered for their determination without modification.

The inhibiting effects of Hg(II) and Ag(I) on ADH were studied in an attempt to determine whether the S_2 emission could be used indirectly to measure the concentration of each of these ions. In these experiments, the yeast ADH solution was first mixed with the solution containing Hg(II) or Ag(I) ions in appropriate concentration. After leaving in ice for 15 min, a solution containing NAD^+, ethanol and Tris buffer at 25°C was added. The mixture was left in a thermostated bath at this temperature for 6 min, when an excess of Ag(I) ions were added to stop the reaction completely. Sodium sulphite was added to react with the acetaldehyde produced, and the remainder of the procedure completed as described above.

The enzyme concentration was chosen so that any trace of inhibitor would reduce its activity. Ethanol and NAD^+ concentrations were such that S_2 emission intensities were readily observable. An incubation time of 15 min gave maximum inhibition of yeast ADH by Hg(II) or Ag(I) ions. Adsorption of these ions on the glassware was minimized by leaving the solutions in contact with the containers for as short a time as possible.

Calibration plots for Hg(II) and Ag(I) were linear from 50–175 and 75–225 pg per 5 μL, respectively. They show that mercury which is thought to bind more strongly to the —SH groups on the enzyme (67) can be determined somewhat more sensitively than silver.

REFERENCES

1. S. L. Bogdanski, M. Burguera, and A. Townshend, *CRC Crit. Rev. Anal. Chem.*, **10**, 185 (1981).
2. R. M. Dagnall, D. J. Smith, K. C. Thompson, and T. S. West, *Analyst.*, **94**, 871 (1969).
3. J. M. S. Butcher and G. F. Kirkbright, *Analyst.*, **103**, 1104 (1978).
4. A. Z. Reis, *Phys. Chem.*, **76**, 560 (1911).
5. A. Fowler and J. S. Badami, *Proc. Roy. Soc. (London)*, **A133**, 325 (1931).
6. R. W. B. Pearse and A. G. Gaydon, *The Identification of Molecular Spectra*, 4th ed., Chapman and Hall, London, 1974.
7. T. M. Sugden, in *Annual Review of Physical Chemistry*, vol. 13, H. Eyring, C. J. Christensen, and H. S. Johnston, Eds., Annual Reviews, Palo Alto, CA, (1962), p. 369.
8. I. H. El-Hag, PhD Thesis, University of Birmingham, Birmingham, UK, 1982.
9. H. P. Broida, H. I. Schiff, and T. M. Sugden, *Trans. Faraday Soc.*, **57**, 259 (1961).
10. A. C. Calokerinos, PhD Thesis, University of Birmingham, Birmingham, UK, 1978.
11. B. E. Buell, *Anal. Chem.*, **34**, 635 (1962).
12. R. Belcher, S. L. Bogdanski, E. Henden, and A. Townshend, *Anal. Chim. Acta*, **113**, 13 (1980).
13. B. Phillipp and H. Dautzenberg, *Faserforsch. Textiltech.*, **19**, 23 (1968).
14. N. V. Sidgwick, *The Organic Chemistry of Nitrogen*, Clarendon Press, Oxford, 1937.
15. J. V. Braun, *Ber.*, **35**, 817 (1902).
16. J. Leonis, *Compt-rend. Lab. Carlsberg. Ser. Chim.*, **26**, 316 (1947–49).
17. R. Belcher, S. L. Bogdanski, and A. Townshend, *Anal. Chim. Acta*, **67**, 1 (1973).
18. S. A. Al-Tamrah, R. Belcher, S. L. Bogdanski, A. C. Calokerinos, and A. Townshend, *Anal. Chim. Acta*, **105**, 433 (1979).
19. I. Z. Al-Zamil, S. A. Sultan, and Y. A. Hassan, *Anal. Chim. Acta*, **239**, 161 (1990).
20. L. S. Snyder and P. Z. Solocins, *Anal. Biochem.*, **65**, 284 (1975).
21. A. F. S. A. Habeeb, *Anal Biochem.*, **14**, 328 (1966).
22. R. B. Freedman and G. K. Radda, *Biochem J.*, **108**, 383 (1968).
23. R. S. Braman, *Anal. Chem.*, **38**, 734 (1966).
24. R. Belcher, S. L. Bogdanski, A. C. Calokerinos, and A. Townshend, *Analyst.*, **102**, 220 (1977).
25. R. Belcher, S. A. Ghonaim, and A. Townshend, *Anal. Chim. Acta*, **71**, 255 (1974).
26. R. Belcher, S. L. Bogdanski, S. A. Ghonaim, and A. Townshend, *Anal. Chim. Acta*, **72**, 183 (1974).
27. R. Belcher, S. L. Bogdanski, A. C. Calokerinos, and A. Townshend, *Analyst.*, **106**, 625 (1981).
28. A. Celik and E. Henden, *Analyst.*, **114**, 563 (1989).
29. E. Henden, *Anal. Chim. Acta*, **173**, 89 (1985).
30. R. Belcher, S. L. Bogdanski, E. Henden, and A. Townshend, *Anal. Chim. Acta*, **92**, 33 (1977).
31. I. M. A. Shakir, S. Y. Atto, and F. H. Hussein, *J. Univ. Kuwait (Sci.)*, **15**, 59 (1988).
32. I. M. A. Shakir, S. Y. Atto, and A. J. Neshteeman, *J. Univ. Kuwait (Sci.)*, **15**, 269 (1988).

REFERENCES

33. I. M. A. Shakir and A. D. Kadhir, *J. Univ. Kuwait (Sci.)*, **16**, 261 (1989).
34. P. R. Hesse, *A Textbook of Chemical Analysis*, Chemical Publishing Company, New York, NY, 1972.
35. J. Mertens, P. Van den Winkel, and D. L. Massart, *Anal. Chem.*, **47**, 522 (1975).
36. I. Z. Al-Zamil and A. Townshend, *Anal. Chim. Acta*, **142**, 151 (1982).
37. T. J. Chow and M. S. Johnstone, *Anal. Chim. Acta*, **27**, 441 (1962).
38. *Annual Book of ASTM Standards*, American Society for Testing and Materials, Easton, PA, 1982, Part 31, Nos. D 1426, D 992, and D 3867.
39. A. Henriksen and A. R. Selmer-Olsen, *Analyst.*, **95**, 514 (1970).
40. International Standards Organisation, Standard Nos. ISO 2918 (1975) and ISO 3091 (1974), International Organisation for Standardisation, Geneva, Switzerland.
41. S. S. Brody and J. E. Chaney, *J. Gas Chromatog.*, **4**, 42 (1966).
42. M. C. Bowman and M. Beroza, *Anal. Chem.*, **40**, 1448 (1968).
43. S. L. Bogdanski, PhD Thesis, University of Birmingham, Birmingham, UK, 1973.
44. R. Belcher, S. L. Bogdanski, D. J. Knowles, and A. Townshend, *Anal. Chim. Acta*, **77**, 53 (1975).
45. O. Osibanjo, PhD Thesis, University of Birmingham, Birmingham, UK, 1976.
46. A. Syty, *Anal. Lett.*, **4**, 531 (1971).
47. D. J. Knowles, P. Marriott, and S. J. E. Slater, *Proc. Anal. Div. Chem. Soc.*, **15**, 62 (1978).
48. P. Krnac, M. Hejtmanek, and B. Polej, *Vysoke skoly chemicko-technologicke v Praze*, H20, 133 (1985).
49. K. Nakajima, K. Ohta, and T. Takada, *Anal. Chim. Acta*, **270**, 247 (1992).
50. A. C. Calokerinos and T. P. Hadjiioannou, *Anal. Chim. Acta*, **157**, 171 (1984).
51. A. C. Calokerinos and T. P. Hadjiioannou, *Anal. Chim. Acta*, **148**, 277 (1983).
52. E. M. Bulewicz and T. M. Sugden, *Proc. Roy. Soc. (London) Ser. A.*, **277**, 143 (1964).
53. O. Osibanjo, A. Townshend, and S. A. Al-Tamrah, *Anal. Chim. Acta*, **162**, 409 (1984).
54. S. A. Al-Tamrah, *J. Coll. Sci. King Saud University*, **15**, 531 (1984).
55. I. H. El-Hag and A. Townshend, *J. Anal. Atom. Spectrosc.*, **1**, 383 (1986).
56. G. Salet, *Compt. Rend.*, **68**, 404 (1869).
57. R. Belcher, S. L. Bogdanski, O. Osibanjo, and A. Townshend, *Anal. Chim. Acta*, **84**, 1 (1976).
58. X. Wang and B. Deng, *Fenxi Huaxue*, **12**, 430 (1984) CA 101:103184.
59. P. Krnac and M. Hejtmanek, *Vysoke skoly chemicko-technologicke v Praze*, H **23**, 53 (1989).
60. R. Belcher, S. L. Bogdanski, M. Burguera, E. Henden, and A. Townshend, *Anal. Chim. Acta*, **100**, 515 (1978).
61. M. J. Cope and A. Townshend, *Anal. Chim. Acta*, **134**, 93 (1982).
62. M. J. Cope, *Anal. Proc. (London)*, **17**, 273 (1980).
63. M. Q. Al-Abachi, R. Belcher, S. L. Bogdanski, and A. Townshend, *Anal. Chim. Acta*, **92**, 293 (1977).
64. T. S. Al-Ghabsha, PhD Thesis, University of Birmingham, Birmingham, UK, 1979.
65. I. Z. Al-Zamil, Y. A. Hassan, and S. M. Sultan, *Anal. Chim. Acta*, **233**, 307 (1990).
66. I. Z. Al-Zamil and A. Townshend, *Anal. Chim. Acta*, **207**, 355 (1988).
67. P. J. Snodgrass, B. L. Vallee, and F. L. Hoch, *J. Biol. Chem.*, **235**, 504 (1960)

CHAPTER

7

HALIDES AND METALS

DAVID A. STILES

Acadia University
Nova Scotia, Canada

7.1	**Introduction**	171
7.2	**Halogens**	171
	7.2.1 Fluorine	172
	7.2.2 Chlorine, Bromine and Iodine	174
7.3	**Metals**	187
7.4	**Conclusions**	192
References		192

7.1 INTRODUCTION

The identification, confirmation and quantification of halogens in a wide variety of analytical environments has been a goal of analytical chemists for many years. With the advent of instrumental techniques, it has become possible to determine these elements in the presence of one another and in the presence of others, at extremely low levels of detection. Molecular emission cavity analysis (MECA) is one such analytical technique that has been developed successfully for the determination of halogens at trace levels.

This technique has also been investigated for its effectiveness in being able to determine trace quantities of metals. However, its ability to outperform methods based on atomic absorption spectrophotometry has been demonstrated in few instances.

The current chapter outlines the development of MECA as an analytical technique for the determination of the halogen elements. It also reviews the use of MECA in determining those metals, not discussed separately in Chapter 5.

7.2 HALOGENS

Historically the determination of fluorine has been considered separately from that of the other three naturally-occurring halogens, because its chemical and

---

Flame Chemiluminescence Analysis by Molecular Emission Cavity Detection
Edited by D. A. Stiles, A. C. Calokerinos and A. Townshend © 1994 John Wiley & Sons Ltd

physical properties are so different. In spectroscopic determinations, chlorine, bromine and iodine are frequently quantified directly by measuring the emission intensity of a metal–halogen species. However, in the case of fluorine this is not possible because metal–fluorine emitting species rarely emit in the ultraviolet-visible part of the spectrum. To measure fluorine it is necessary to make an indirect determination by first converting the element to a fluorine-containing compound and then measuring an absorption or emission intensity due to one of the other elements present. This is also true for the MECA determination of halogens.

7.2.1 Fluorine

Although Gutsche et al. have reported the determination of fluorine using Ca–F emission bands (1), this study was concerned primarily with the gas chromatographic detection of organofluorine compounds and the development of a gas chromatographic detection system specific to fluorine. The same laboratory (2) also reported that the InF species emitted under conditions similar to those found inside the MECA cavity. However, neither CaF nor InF emitting species have been used in the MECA determination of fluorine.

The method found to be most satisfactory for the MECA determination of fluorine involves the conversion of fluorine to volatile silicon tetrafluoride which is then quantified through the emission intensity of Si–O bands in an oxy-cavity (3). This method, therefore, involves the indirect determination of fluorine and is, in fact, similar to that reported for the direct determination of silicon (4) (see Chapter 5).

In fluorine determinations by MECA, sodium silicate and fluoride as aqueous solutions were pipetted in known quantities into a PTFE reaction vessel fitted with PTFE delivery inlet and outlet ports (Figure 5.6). The mixture was evaporated to dryness at 110°C, cooled and then connected to a nitrogen purge and cavity. Concentrated sulphuric acid was then injected onto the dried sample through a septum and the reaction vessel heated at 180°C for 5 min. Silicon tetrafluoride, formed in the reaction, was then purged into the cavity through the delivery tubes. In order to prevent condensation of water and subsequent reaction and loss of silicon tetrafluoride in these tubes, they were wrapped in heating tape maintained at a temperature just above 100°C. The cavity used was a water-cooled oxy-cavity constructed of aluminium.

During the course of measuring the emission intensity of the Si–O band (λ_{max} = 540 nm), gas flow rates were maintained at: H_2, 2 L min^{-1}; N_2, 5.0 L min^{-1} (flame); O_2, 110 mL min^{-1} (purge); N_2, 80 mL min^{-1} (purge). Hotter flames generally gave higher emission intensities, while cooler flames decreased the background noise. The upper limit for hydrogen flow rate and, therefore, the temperature of the cavity was determined by its resistance to thermal deformation.

As in the case of silicon determinations (see Chapter 5), experiments using different quantities of silicon and fluorine, and other forms of silicon-containing volatiles, confirmed that the product of the reaction in the PTFE vessel was silicon tetrafluoride, and that the white emission observed in the flame was due to Si–O emitting species. No emission was observed when silicon was treated in the recommended fashion in the absence of fluorine.

Linear calibration graphs were obtained for fluorine in the range 10–200 μg, with a maximum coefficient of variation of 6% at the lower concentrations. The detection limit was 5 μg fluorine.

Twenty-fold additions of Fe(III), Cr(III), Cu(II), Ni(II), Co(II), Pb(II) and Cd(II) as their nitrates, sulphates and phosphates caused no interference in the determination of 50 μg of fluorine. Strong acids interfered if they were added to the reaction prior to the evaporation step because they produced volatile HF which disappeared through evaporation during the drying process. This potential interference was eliminated by using a reaction mixture pH greater than 7.

Ammonium salts were another potential source of interference. Normally they would be converted to volatile ammonia under basic conditions. If they were present with silicates, therefore, the possibility exists that ammonia and silicon tetrafluoride would be liberated in the reaction vessel at the same time. Ammonia is known to produce a white emission band of NO–O in the oxy-cavity (5) and it is very likely that this would interfere spectrally with that of Si–O. The potential interference was eliminated altogether in the drying step where the ammonia was evaporated, leaving the silicon-containing material behind.

Elements such as arsenic and boron which form volatile fluorides under the conditions present in the reaction vessel (6) interfered by (a) using up fluorine, and (b) giving alternative emissions (7) which interfered in the measurement of the Si–O emission intensity if the slit width of the monochromator was greater than 0.5 mm. These unwanted emission bands could be partially eliminated as a spectral interference by centring the monochromator at 580 nm rather than 540 nm and using a slit width of only 0.2 mm. Under these conditions the allowable amounts of arsenic and boron were 5 μg and 40 μg respectively for the determination of 50 μg of fluorine.

Experiments conducted using arsenic or boron as the volatile fluorine-generating agent showed that these two elements could be used for the determination of fluoride with reasonable sensitivity and precision. Comparative information on the determination of fluoride with arsenic, boron and silicon fluorides is given in Table 7.1.

The silicon tetrafluoride method was used to determine the amount of fluoride in toothpaste and canal water (3) and the results compared with those using the alizarin fluorine blue (AFB) method (8–10). In each case the results obtained using MECA were comparable to those using the

Table 7.1 Comparison of AsF_3, SiF_4, and BF_3 for the Indirect Determination of Fluoride by MECA using the Recommended Conditions for SiF_4 Generation

Product	B.Pt (°C)	λ_{max} (nm)	Linear Range (µg)	µg F	Relative Standard Deviation (%)	Detection Limit (µg)[1]
AsF_3	−62	490	8–200	25	4.0	1.6
SiF_4	−86	540	20–225	50	4.2	5
BF_3	−100	518	50–300	100	3.3	10

[1] Defined as the amount of fluoride producing a signal twice the background deviation.

more classical determination (Table 7.2) and in the case of the toothpaste had the advantage of not being susceptible to phosphate interference.

7.2.2 Chlorine, Bromine and Iodine

The determination of these species by flame methods is relatively old, dating back to the discovery that copper halides give a particularly intense emission when introduced into a flame. This provided the basis for the Beilstein method for halogen identification (11). However, it was non-specific because the observed emissions were due mainly to oxygen- and hydrogen-containing species such as CuO, CuOH and CuH, and could be obtained by introducing pseudohalides such as cyanide and thiocyanate into the flame. Only weak emissions were obtained from the copper halides and these were too weak to be of any use in analytical determinations.

However, over succeeding years other metals were found to give more intense metal halide emissions, indium being particularly useful for analytical purposes because of its relatively strong indium–halogen bond (12,13).

Table 7.2. Comparative Determination of Fluoride in Water and Toothpaste Using MECA and Absorption Spectrophotometry (AFB Method)

Sample		Amount of Fluoride	
		MECA	AFB
Water	Tap	0.92	0.93
(mg F mL^{-1})	Canal	0.65	0.63
Toothpaste	Crest (SnF_2)	0.098	0.095
(% F)	Colgate $(FPO_3)^{2-}$	0.106	0.010[1]
	MacLeans $(FPO_3)^{2-}$	0.103	0.010[1]

[1] Phosphate interferes seriously.

Two methods have been used to stimulate the metal–halogen emission. In the first, indium(III) nitrate was added to the halogen-containing solution which was then aspirated directly into a hydrogen diffusion flame. In the alternative method, the indium was placed in the flame as a screen or indium-plated copper tube and the halogen-containing solution aspirated as before.

The indium halide emissions for chloride and bromide were found to be sufficiently stable and strong that they were incorporated into the design of a halogen detector for gas chromatographic studies (14). Examples where the indium halide emission was used for practical purposes was in the determination of chlorine and bromine in a number of agricultural commodities (15).

The first reports on the use of indium in MECA studies consisted of an evaluation of two techniques. In one, indium(III) nitrate was added to the halide in aqueous solution and the resulting mixture injected as aliquots into a stainless-steel cavity. The second, and more satisfactory method, involved injection of an aliquot of halogen-containing solution into a stainless-steel cavity which had been previously coated with a layer of indium (16,17). So long as the nitrogen-diluted hydrogen diffusion flame which heated the cavity and caused indium halide emission was extinguished as soon as the InX emission had subsided, reproducible results free of any interference from other emissions such as In_2 were obtained. A typical example of an indium-lined cavity halide emission is shown in Figure 7.1.

Studies showed that maximal InX emission occurred when the concentration of halide to indium was in the ratio of three to one but that an excess of indium was required for reproducible results. Belcher *et al.* (17) reported that the method was satisfactory for the determination of nanogram quantities of chloride, bromide and iodide with limits of detection equal to 2.5 ng for Cl^- and Br^- and 50 ng for I^-. Calibration curves were linear within the ranges 2.5–350 ng (Cl^-); 2.5–40 ng (Br^-) and non-linear for iodide.

In the same study, it was found that ions which formed a less volatile compound with either the indium or halogen, through a metathesis reaction, interfered, for example

$$In^{3+} + PO_4^{3-} \longrightarrow InPO_4$$

$$Na^+ + X^- \longrightarrow NaX$$

Anionic interferences were, to a large extent, eliminated by using excess indium, a condition fully met in the indium-lined cavity. Cationic interferences were not resolved, however, since sodium, calcium and other metals would have required a stoichiometric excess of the ion being determined in order to eliminate their effect. Clearly this would not be feasible.

One of the problems in determining halogens by MECA compared to the determination of other non-metals is that the halogen is always converted first to a common metal halide species, usually InX, and that its t_m does not

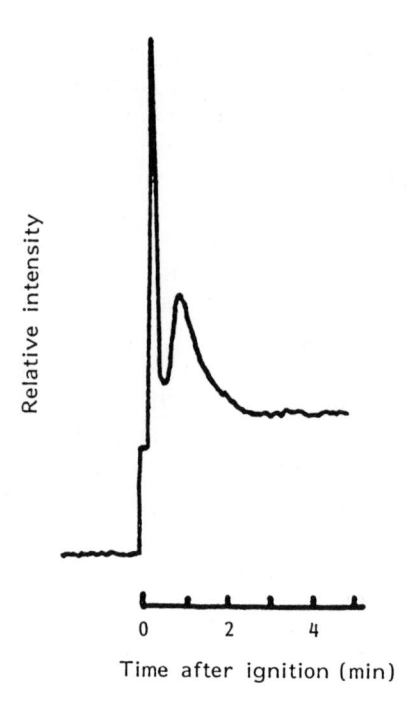

Figure 7.1. Responses from InCl and In$_2$ in an indium-lined stainless-steel MECA cavity showing temporal resolution of the emissions.

change significantly enough from one analyte to another for it to be of any analytical usefulness. For this reason, halogen determinations have been accomplished either by first converting the analyte to a common halogen-containing species, e.g. HCl, and determining the total halogen, i.e. chlorine, or by separation of analytical components prior to their introduction into the cavity. The former technique was used successfully for determining nanogram quantities of *p,p'*-DDT, γ-BHC and Heptachlor (18), ethylene dibromide (EDB) (19,20), ethylene chlorobromide, bromochloropropene and ethylene dichloride (20), and polychlorinated biphenyls (PCBs) (21). In each case the organohalogen compound was first converted to HCl, HBr, NH$_4$Cl or NH$_4$Br prior to quantification. Narine *et al.* (18) and Abdel-Kader (19) used a Schöniger oxygen combustion flask to convert the analyte to HCl or HBr. In these experiments, soil which had been fortified with organohalogen material was extracted into a mixture of either acetone–hexane or acetone–isooctane before being decomposed. The products of combustion were then dissolved in a minimal amount of water before being injected into an indium-lined cavity for quantification.

Interference from potassium present in the combustion paper was insignificant at the concentrations of analyte used. However, the amount of paper used relative to the amount of analyte being decomposed is important because with a reduced amount of paper, a carbonaceous reducing agent, both hydrogen halide and pure halogen are formed, the latter being dissolved less readily than the halide. This becomes an important factor as the amount of analyte is increased (22).

In the case of EDB, the Soxhlet extraction procedure used by Narine et al. (18) for relatively low vapour pressure pesticides was inappropriate because of the greater volatility of EDB. Hence a procedure reported by Malone (23) was modified slightly to extract the analyte from the soil. However, in all other respects, the experimental method used was the same as that used in the chlorine-containing pesticide determination. Both organochlorine pesticides and EDB were determined at the nanogram level with percentage recoveries above 94% for different soil types.

Since bromine and chlorine interfere with one another spectrally at the fairly wide collimator slit widths used in MECA spectrometers, it became necessary to develop a different procedure in those cases where either the analyte contained both elements, or where mixtures containing both halogens were being analysed. Abdel-Kader et al. (20) were able to resolve this dilemma by dissolving the products of Schöniger flask combustion in ammonia rather than water. The ammonium halides thus formed were separated by thin layer chromatography and then quantified separately by MECA.

A further modification to the combustion procedure was followed by Persaud et al. (21) who used a combustion tube technique rather than an oxygen flask. Previous investigators had reported that the combustion tube technique was suitable for the microcoulometric determination of halogens (24,25). This method was investigated for the determination of PCBs at the microgram level as total chlorine in fortified soil. The effects of a variety of common non-metallic and cationic potential interferences were studied. In most instances, chlorine recoveries were over 90%. Interferences were observed from sulphur-containing species and the alkali metal or alkaline earth cations. Carbon, nitrogen, phosphorus, iron and aluminium caused no noticeable interference at microgram levels of fortification.

An attempt was made to eliminate the cationic interfering species by treating samples fortified with sodium, potassium, calcium and magnesium with EDTA before they were combusted. No noticeable increase in percentage recovery of chlorine was observed. In another attempt to remove these interferences, the organochlorine compound was extracted from the fortified soil into a mixture of acetone–hexane, after which it was combusted. This technique appeared to separate alkali metal cations from the organochlorine material and increased the recovery of this element from approximately 60 to 90%. No such increase in percentage recovery was noted for the alkaline earth elements.

In the case of the sulphur interference, use was made of the fact that the various sulphur-containing emissions have a slightly different t_m from that of InCl. At normal recorder chart speeds (1.25 cm min^{-1}), the sulphur peaks

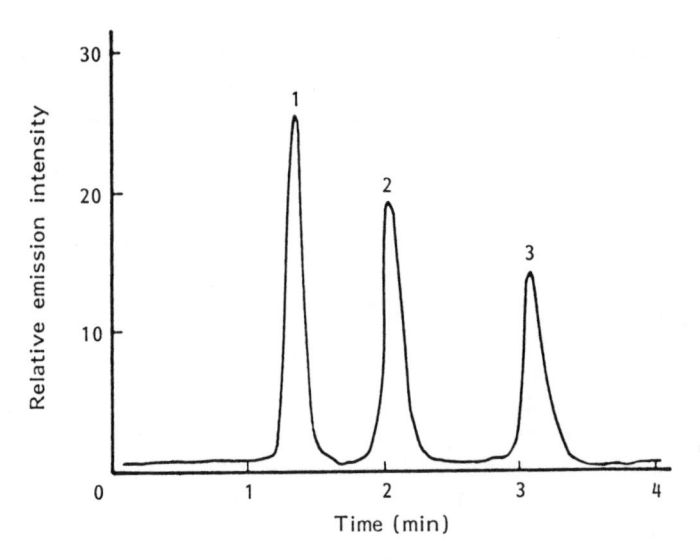

Figure 7.2. GC–MECA response for 10 mg L^{-1} Cl as (1) CH$_2$Cl$_2$; (2) CHCl$_3$; (3) CCl$_4$.

overlapped with that of InCl and no resolution of the components was possible. However, at a chart speed of 12.5 cm min^{-1} each sulphur-containing species appeared at a t_m well-resolved from that of InCl, thus allowing independent quantification of the halogen.

A different approach to organohalogen detection by MECA was taken by Burguera et al. (26,27) who attached the cavity to the outlet port of a gas chromatograph. Separation of mixtures took place within the chromatograph so that it was possible to determine individual components rather than simply a value for total halogen in the mixture (see Chapter 3).

A study of the effect of the carrier gas flow rate on the emission intensity of methylene dichloride, chloroform and carbon tetrachloride showed that a flow rate of 20 mL min^{-1} gave excellent resolution of individual chromatographic peaks (Figure 7.2). This carrier gas flow rate ensured that the solvent, ethanol, eluted well after the components of interest and did not quench InCl emissions.

Calibration curves for each of the compounds studied were linear up to 60 mg L^{-1} chlorine. Above this concentration the graph flattened out due to chlorine saturation of the detector. The detection limit for chloroform measured as twice the noise level was 1.5 ng of chlorine (0.23 mg L^{-1}) with a relative standard deviation of 2.4% for six 5 μL replicates containing 5 mg L^{-1} chlorine.

In the second study (27), a water-cooled stainless-steel cavity was used, giving more temperature control. MECA parameters were optimized for chlorine-, bromine- and iodine-containing organohalogen compounds. Each of them was vaporized in a PTFE generator and the vapour carried to the cavity in a stream of nitrogen.

The emission spectra of each compound studied showed that these were the same for all compounds containing the same halogen and that the intensity of the appropriate InX emission was directly proportional to the amount of halogen vaporised. Water cooling of the cavity affected the emission intensity, however.

At lower flow rates, the emission intensity decreased as the cavity temperature increased. However, for chlorine- and bromine-containing compounds it was constant for a water flow rate of between 80 and 300 mL min^{-1}, so that in all further studies on these two types of compounds a flow rate of 150 mL min^{-1} was used.

For iodine-containing compounds where the concentration of iodine was determined from the difference in InI and In emissions at 410 nm, the InI emission intensity increased at the expense of that due to In as the water flow rate increased. Thus for studies involving iodine, a water flow rate of 400 mL min^{-1} was used, this being the maximum carrying capacity of the equipment.

A spectral interference of bromine on chlorine in the absence of GC separation was observed at relatively high bromine concentrations and at oven temperatures above 100°C. This interference was eliminated successfully by keeping the oven temperature at 100°C and maintaining the concentration of bromine below 400 mg L^{-1}. No corresponding spectral interference of chlorine on the bromine emission at 376 nm was observed at the concentrations studied. Iodine had no significant effect on the emission intensity of either chlorine or bromine.

In MECA–GC studies, ethanol, benzene, n-hexane and acetone were investigated for use as solvents. Although both acetone and ethanol were considered suitable, only the latter was used in subsequent work. Benzene and n-hexane were unsuitable because they quenched InX emissions. Samples containing mixtures of organochlorine, organobromine and organoiodine compounds were separated by gas chromatography and then detected individually by MECA.

An interesting application of MECA to the determination of phenol in lake waters was reported by Osibanjo and Ajayi (28). In their study, lake water was first filtered, acidified with glacial acetic acid, and then treated with a small stoichiometric excess of bromine. The tribromophenol precipitate was filtered, washed until it was free of bromine and then dissolved in ethanol. Aliquots of this solution were injected into an indium-lined MECA cavity and the emission intensity measured at 376 nm.

Standards of pure phenol were used for calibration and the results were confirmed using the method of standard additions. The percentage recovery of phenol was found to be over 98% with a reported water content of 9×10^{-5} mol L^{-1} (0.85 mg L^{-1}) phenol. No interference from nitrate, phosphate or sulphate was observed. Chlorine and iodine, while posing potential interferences through the formation of trichloro- and triiodo-phenols were

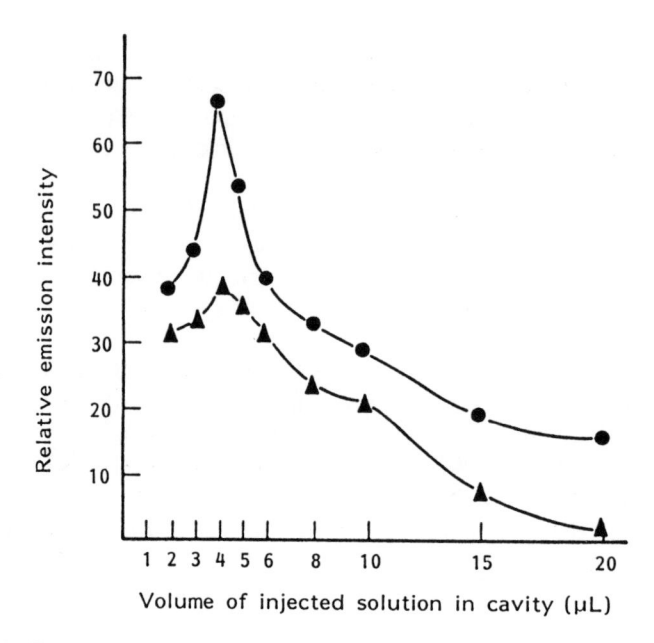

Figure 7.3. Effect of injected volume of GaCl and GaBr on their MECA emission intensities: (●) GaBr; (▲) GaCl.

not considered a problem in this analysis because the interfering ions occurred at very low levels in natural waters.

Kassir et al. (29) and Osibanjo et al. (30) have extended these investigations to the determination of trace levels of halogen as gallium and copper halides respectively. Kassir et al. were also able to utilize their procedure for the determination of gallium metal. The investigations using gallium halides were noteworthy because the authors found that argon–hydrogen was more sensitive than nitrogen-hydrogen as flame gas. In addition, they confirmed that the best metal-to-halide concentration ratio for the determination of bromide and chloride was 1:3 (16).

Kassir et al. (29) used the GaCl line at 338 nm and the GaBr line at 350 nm in their study. The GaCl line was chosen because it was inherently more intense (Figure 7.3); that for GaBr because an enhancement in intensity was observed on the addition of hydroiodic acid to samples (Figure 7.4). No spectral interference from gallium iodide was observed. It is also interesting to note that in this study the cavity giving best results was made of carbon rather than stainless-steel or other material. In this situation, it was necessary to add the gallium to the halide in aqueous homogeneous solution before injection into the cavity. This procedure is in direct contrast to most of the MECA halogen determinations where the cavity is lined with metal, or a metallic grid is used.

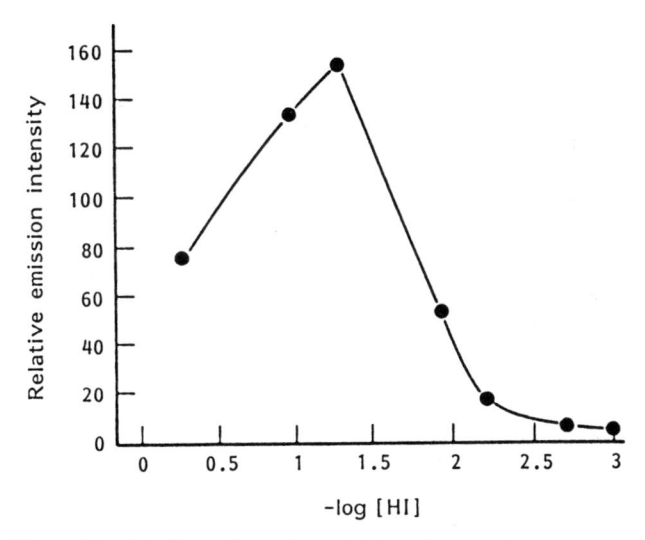

Figure 7.4. Effect of hydriodic acid on the MECA emission of GaBr.

Using the iodide-enhancing technique, bromide gave a linear calibration graph up to 60 ng for standards, the detection limit being 0.5 ng μL^{-1} with a 6.9% relative standard deviation. For samples containing an unknown quantity of bromide where strict control of a 1:3 metal:halide ratio was impossible, gallium(III) nitrate concentration in excess of this stoichiometric ratio was maintained. In this case the detection limit increased to 40 ng μL^{-1}. In chloride determinations, the detection limit in the presence of excess gallium was 50 ng μL^{-1} with a linear calibration range of between 0 and 1200 ng chloride.

Table 7.3. Effect of Various Interfering Ions on the MECA Emission Intensity of GaBr

Ion	Change in Intensity (%)			Ion	Change in Intensity (%)		
	$1\times^1$	$2\times^1$	$5\times^1$		$1\times^1$	$2\times^1$	$5\times^1$
Na^+	−78	−90	−92	Fe^{3+}	−35	−64	−70
K^+	−53	−64	−73	Al^{3+}	−46	−50	−55
Mn^{2+}	−85	−91	−94	In^{3+}	+>100	+>100	+>100
Cu^{2+}	−82	−87	−94	F^-	−3	−10	−58
Cd^{2+}	−53	−70	−32	Cl^-	−27	−62	−63
Ni^{2+}	−67	−80	−86	NO_3^-	+1	+2	+3
Co^{2+}	−74	−87	−93	SO_4^{2-}	+12	+>100	+>100
Zn^{2+}	−45	−83	−93	PO_4^{3-}	−43	−78	−86
Mg^{2+}	−8	−22	−44				

[1] Weight excess.

Table 7.4 Effect of Flame Composition in Different Flame Systems on MECA Emissions from Copper Halides Using a Copper Cavity

Halide species as CuX	Cavity Position[1]			Flame Composition			Types of Emission Observed (colour)
	CB (mm)	CMS (mm)	HD (mm)	Nitrogen (L min^{-1})	Air (L min^{-1})	Hydrogen (L min^{-1})	
X = Cl$^-$	16	60	8	8	–	3	Light green → green-white → white
				–	10	5	Intense green
				6	4	2	Light green → green-white → white
X = Br$^-$	16	60	8	2	–	5	Intense light blue → blue-white
				–	9	5	Very intense blue
				4	7.5	4	Intense light-blue → blue-white
X = I$^-$	16	60	8	5	–	4	Rich green
				–	10	5	Deep green
				5	7	5	Deep green

[1] CB = cavity–burner distance, CMS = cavity–monochromator slit distance, and HD = horizontal distance of the cavity into the flame relative to the flame edge.

A variety of interferences were investigated (Table 7.3). Most caused a decrease in halide emission intensity, although indium and nitrate caused enhancement.

The study reported by Osibanjo et al. (30) is a comprehensive investigation of several experimental parameters on copper halide emissions in a MECA cavity and the use of copper bromide, chloride and iodide emissions to determine quantities of individual halogens. It is an important study from a spectroscopic point of view because it provides more useful information on copper-containing species present in the relatively cool MECA cavity, and confirms that CuX emitting species are dominant in that environment.

In this investigation, a copper MECA cavity was used and samples containing halogenated species as aqueous solutions were injected directly into it. Using different mixtures of hydrogen, nitrogen and air for the flame, a variety of emissions were observed (Table 7.4). However, the most noteworthy feature of these spectra was that the intensity maxima for CuCl and CuBr occurred at 532 nm and 494 nm rather than at 488 nm as reported previously (31). The intensity maximum for CuI emissions occurred at the expected value of 510 nm (31). A complete documentation of cavity emission features for each flame gas mixture studied was presented (Table 7.5).

A mixture of gases containing hydrogen, nitrogen and air suppressed the formation of unwanted alternative copper-containing species such as CuH, CuO and CuOH and was therefore used in all experiments to produce calibration graphs and to determine detection limits. Linear calibration graphs were obtained for bromide from 1×10^{-3} to 6×10^{-3} mol L^{-1}, chloride from 3×10^{-3} to 1×10^{-2} mol L^{-1} and iodide from 1×10^{-3} to 3×10^{-2} mol L^{-1}, using emissions at 434 nm, 532 nm and 510 nm, respectively. Detection limits were as follows: bromide, 1×10^{-3} mol L^{-1}; chloride, 3×10^{-3} mol L^{-1}; iodide, 1×10^{-3} mol L^{-1}, with the coefficient of variation for ten replicates of Br^-, Cl^- and I^- being 6.5, 4.8 and 7.2%, respectively.

Sodium chloride and potassium chloride, which are well-known to suppress InCl emission intensities (17), gave no emission of their own in the copper cavity but depressed emissions from copper halides. Other potential chemical interferences were eliminated through appropriate chemical reactions as indicated in the literature (6,32).

Another interesting point to emerge from this investigation was that the t_m values for the CuBr, CuCl and CuI emitting species depended on flame composition. With the mixture of hydrogen, nitrogen and air used in most of the work reported, the t_m values were: CuBr, 27 s; CuCl, 73 s; CuI, 36 s. This suggests the possibility of determining the components of binary mixtures containing these halogens without a pre-separation step, and is in direct contrast to all previous indicators, as noted in the review by Ajlic and Stupar (33).

Table 7.5 Types of MECA Emissions Observed from Copper Halides Using a Copper Cavity

Halide Species as CuX	Gas Flame Combination	Colour of Emission	Remarks
X = Cl⁻	Hydrogen only	Faint light green	Takes some time to appear and fills the inside of the cavity completely
	Nitrogen–hydrogen (i) Low flow rate of N_2	Light green	N_2 flow rate between 1.0 and 5.0 L min^{-1}
	(ii) Medium flow rate of N_2 (<9.0 L min^{-1})	Green-white	N_2 flow rate between 6.0 and 8.5 L min^{-1}
	(iii) High flow rate of N_2 (>9.0 L min^{-1})	Pure white	Faint light-green emission at the mouth of the cavity vanishes leaving a white emission confined to the inside of the cavity. No white emission
	Air–hydrogen	Intense green	
	Nitrogen–hydrogen (i) Low flow rate of N_2	Light green	N_2 flow rate between 1.0 and 4.0 L min^{-1}
	(ii) Medium flow rate of N_2 (<9.0 L min^{-1})	Green-white	N_2 flow rate between 4.5 and 8.5 L min^{-1}
	(iii) High flow rate of N_2 (>9.0 L min^{-1})	White	White emission obtained but the flame is not as transparent as for the nitrogen–hydrogen flame system. White emission was not as clearly defined as for the nitrogen–hydrogen flame system

$X = Br^-$	Hydrogen only	Rich blue	Takes some time to appear and fills the inside of the cavity completely
	Nitrogen–hydrogen (i) Low flow rate of N_2	Intense light blue	N_2 flow rate between 1.0 and 5.0 L min^{-1}
	(ii) High flow rate of N_2 (>7.0 L min^{-1})	Blue-white	From a flow rate of 7.0 L min^{-1} a blue-white emission is observed that persists as the N_2 flow rate increases
	Air–hydrogen	Very intense blue	No blue-white emission whatever the flow rate of air
	Nitrogen–hydrogen (i) Low flow rate of N_2	Intense light blue	N_2 flow rate between 1.0 and 5.0 L min^{-1}
	(ii) High flow rate of N_2 (>7.0 L min^{-1})	Blue-white	Blue-white emission that persists as the N_2 flow rate increases
$X = I^-$	Hydrogen only or nitrogen–hydrogen	Rich green	Takes some time to appear when H_2 is used, but the emission completely fills the inside of the cavity. With increased N_2 flow rates, a rich green emission is confined to the inside of the cavity, also making the flame transparent

Table 7.5 (*continued*)

Halide Species as CuX	Gas Flame Combination	Colour of Emission	Remarks
	Air–hydrogen or nitrogen–hydrogen		
	(i) Low flow rate of air	Deep green	Deep green emission: deeper than that produced for CuCl under the same system and distinct from it for an air flow rate up to 7.0 L min^{-1} in the nitrogen–air–hydrogen flame
	(ii) High flow rate of air	No emission	An increase in air flow rate beyond 8.0 L min^{-1} quenches the emission (low N$_2$ flow)

7.3 METALS

The impetus towards finding analytical methods for determining trace amounts of metals has, in large measure, been directed towards specific applications. This is because atomic absorption spectrophotometry is a well-established technique that is sensitive, reliable and has general application to trace metal determination. It is not surprising, then, that applications of MECA to this area of research have been comparatively limited.

The first reported use of MECA for the determination of metals other than those such as tin and germanium which form volatile hydrides, was by Belcher *et al.* (34) who investigated emissions from manganese(II), cobalt(II), nickel(II), lead(II), cadmium(II), mercury(II), indium(III) and iron(II). The MECA cavity was constructed of stainless-steel and samples were injected into this as aqueous solutions prior to it being heated in a hydrogen–nitrogen–air flame. Spectra of emitting species were recorded from both the MECA cavity and from aspirated solution. The only elements to produce MECA emissions suitable for trace analytical determinations were indium, which has been discussed in connection with halogen analyses, and cadmium. Mercury(II) halides gave no detectable response.

In the case of manganese(II) halides, elements of general similarity were noted between MECA and aspiration spectra, with the majority of identifiable MECA peaks being ascribed to Mn–X emissions. The manganese(II) iodide MECA spectrum had less structure and was much more intense than the aspiration spectrum between 450 and 575 nm (Figure 7.5).

Cobalt(II) halides gave MECA spectra which differed markedly from their aspiration counterparts. With MECA, each halide gave an intense broad

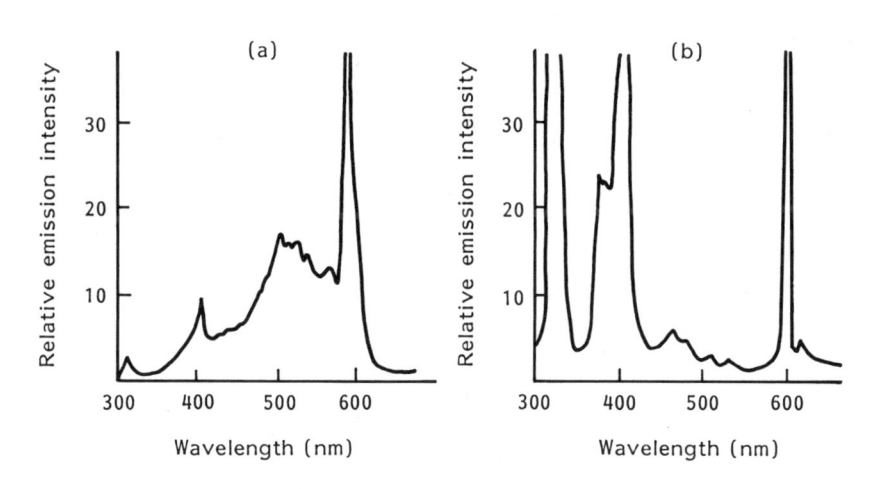

Figure 7.5. Spectra of MnI_2 by (a) MECA, (b) aspiration.

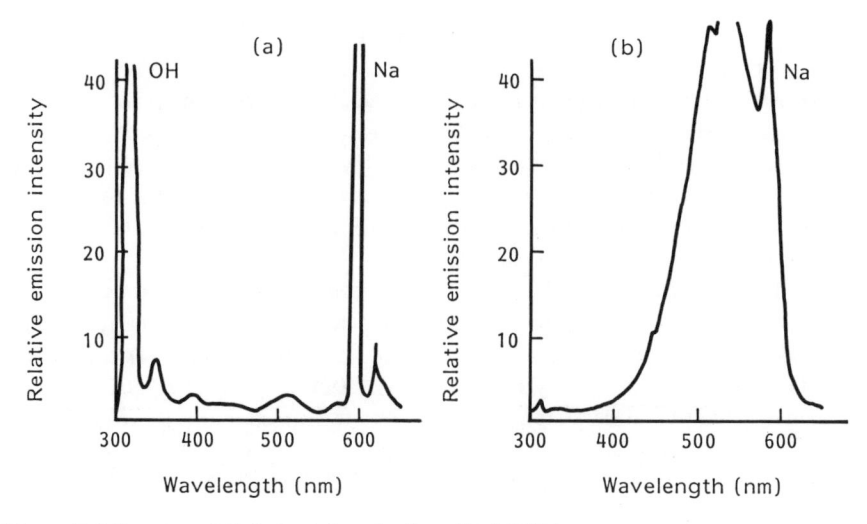

Figure 7.6. Spectra of CoI_2 by (a) aspiration, (b) MECA.

band emission which was considered likely to be due to the appropriate Co–X emitting species. The aspiration spectra, on the other hand, had as their major component the emission from cobalt atoms (Figure 7.6).

The only nickel(II) halide to give a MECA spectrum which was both structured and strong enough to be recorded was the bromide. This was ascribed to a nickel–oxygen rather than a nickel–bromine emission. The chloride gave essentially no MECA spectrum while that of the iodide was described as 'broad and featureless'.

With lead halides, all three gave MECA spectra which appeared to be a superposition of emissions from metal halide and metal oxide to varying degrees. The aspiration spectra tended to be of lower intensity and less structured (Figure 7.7).

Iron(II) chloride gave a MECA spectrum which contained components of Fe–Cl, Fe and FeOH species. No mention was made of MECA spectra obtained from the corresponding bromide and iodide, but the aspiration spectra of chloride and iodide were reported as being due primarily to Fe, FeOH and FeO emissions.

The difference between the aspiration and MECA spectra of the various elements studied was explained on the basis of the temperature at which the emitting species formed and the relative strengths of the metal–halogen bonds. Elements such as indium, which are well-known to have relatively strong M–X bonds, provided an intense enough emission for it to be used as the basis of trace metal analyses; those such as cadmium and mercury, for which the M–X bond is weak, did not give any observable metal–halogen emission. Elements of intermediate metal–halogen bond strength gave weak

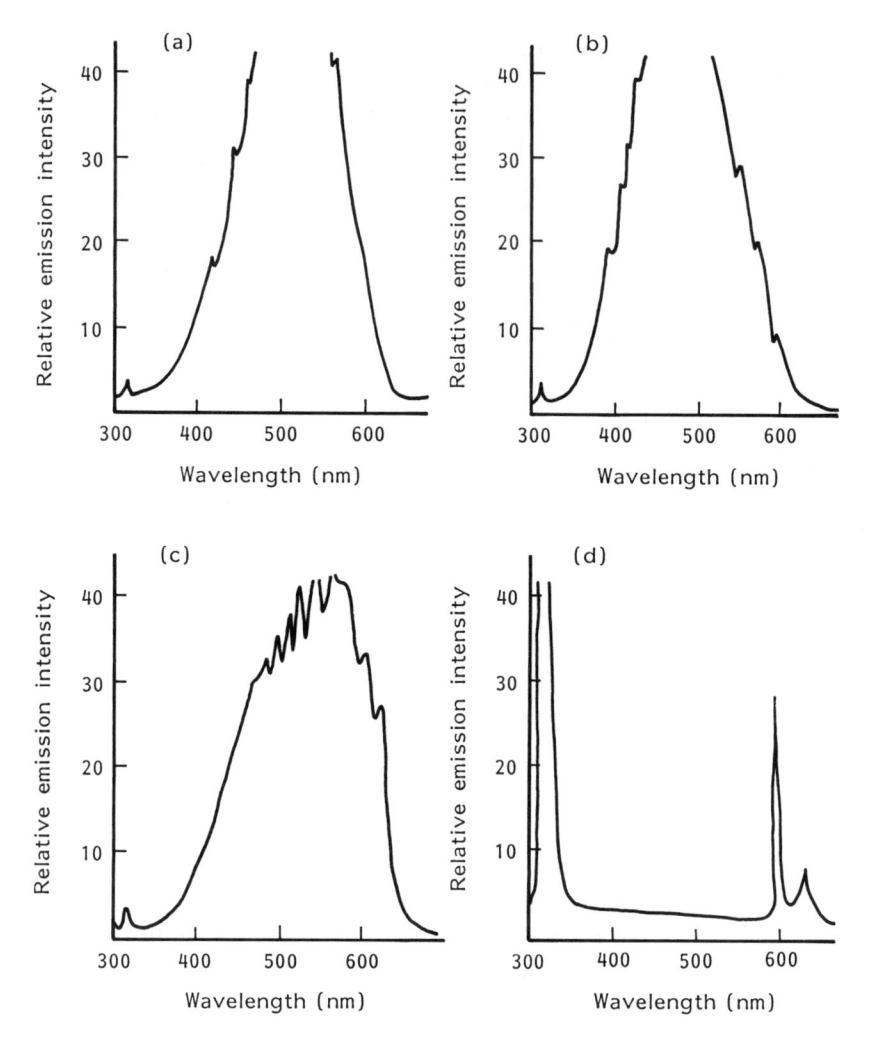

Figure 7.7. Spectra of (a) PbBr$_2$, (b) PbCl$_2$, and (c) PbI$_2$ by MECA, and (d) PbI$_2$ by aspiration.

M–X emissions, confined mainly to the interior of the cavity where the temperature was cooler than in the flame itself and where the Salet effect was most likely to occur.

The determination of trace amounts of cadmium (36) and thallium (37) by MECA provides examples where atomic emission lines rather than those from molecular metal–halogen species have been utilized. For cadmium the line at 326.1 nm was used while in the case of thallium the 377.5 nm line was selected.

In the cadmium study, cavities made of stainless-steel, aluminium and pyrolytic carbon were investigated. Samples of the appropriate cadmium salt were injected into the MECA cavity at room temperature, which was then rotated into a hydrogen–nitrogen–air flame. The emission intensity of the cadmium atomic line was then measured. In these investigations, stainless-steel and carbon cavities were found to be most effective.

Optimum conditions of the flame for observing cadmium emissions were when the flame was fuel-rich. When the percentage of oxygen in the flame was increased, the cadmium signal was diminished, although not with a concomitant increase in Cd–O emitting species.

Interferences were observed from those anions such as silicate, sulphate and phosphate that form thermally stable compounds with cadmium. However these interferences were partially resolved by using a stainless-steel cavity insulated from the cavity holder by a ceramic ring. This had the effect of heating up the cavity rapidly so that it reached operating temperature more quickly and did not allow time for the formation of refractory compounds. An even more satisfactory method for reducing the effect of anionic interferences was to treat all cadmium solutions with sulphuric acid. In this case all cadmium was converted to a compound having a common anion, namely the sulphate.

Most metals were found to decrease the emission intensity of cadmium, although copper had no effect and cobalt caused an enhancement. The most satisfactory method for reducing the effect of these interferences was again to use a cavity thermally insulated from the cavity holder. In this way, 50 ng of cadmium could be determined without interference from a 50-fold weight excess of lithium, sodium, potassium, barium, calcium, nickel, lead or zinc, a 25-fold weight excess of cobalt, manganese and mercury, a five-fold weight excess of tin and magnesium, and a two-fold weight excess of iron and chromium.

Thallium was determined in an oxy-cavity using the acetate and carbonate salts dissolved in either ammonium bromide or ammonium iodide (37). Some difference in emission intensity was noted between the two starting materials, the acetate being more sensitive than the carbonate for a given quantity of analyte. This was attributed to a difference in bond strength and volatility of the respective thallium salts. The calibration curve was linear from 25 to 200 ng with a detection limit of 5 ng thallium in 5 μL of solution. The relative standard deviation was 5% for seven replicates containing 5 μg thallium.

An interesting example of where an interference was utilized positively was demonstrated in the case of potassium ions which enhanced the thallium MECA emission through ionization suppression. When potassium nitrate at a concentration of 1 mg mL^{-1} was added to standard thallium solutions, the emission intensity for that element was increased by 45% (Figure 7.8).

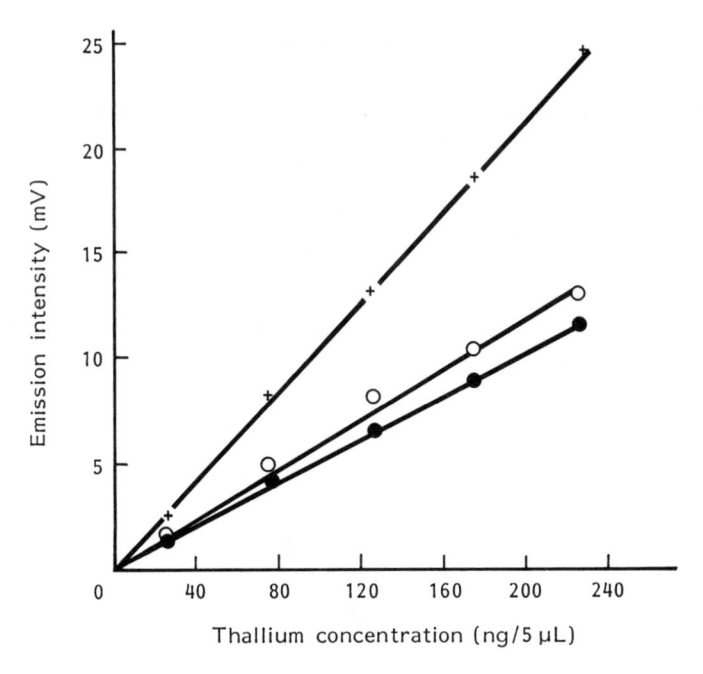

Figure 7.8. Calibration graphs for the MECA determination of thallium: (●) as its carbonate; (○) as its acetate; (×) as its acetate with the addition of potassium nitrate.

Very few other ions caused an interference in thallium MECA determinations. Sulphate and phosphate, which form thermally stable compounds with thallium, as well as manganese(II) and gallium(III), were not important interferences so long as the flame was maintained at a sufficiently high enough temperature. Arsenic(III) and antimony(III) enhanced the thallium emission by forming oxide emitting species in the oxy-cavity.

Gallium, like indium (34), has been determined from its halide emissions in the MECA cavity, although the sensitivity for gallium is much less than for indium. Using carbon or stainless-steel MECA cavities, Al-Tamrah et al. (37) were able to stimulate only bromide and iodide emissions. However, in subsequent work, Kassir et al. (29) observed gallium chloride emissions as well. According to Al-Tamrah et al. the iodide was much more sensitive than the bromide and was therefore used in gallium determinations. They reported that they could determine between 5 and 50 ng of gallium with a detection limit of 0.5 ng per 5 μL of solution. A relative standard deviation of 2.5% was obtained from 10 replicates containing 0.1 μg of the metal.

Kassir and co-workers (29) reported that of the two halides, gallium bromide gave better sensitivity than gallium chloride. However, the range of concentrations for which gallium could be determined was greater for chloride detection (< 400 ng μL^{-1}) than for bromide (< 60 ng μL^{-1}).

7.4 CONCLUSIONS

Although conventional MECA may be used successfully in determining 'total halogen' at trace levels, it is of limited use where specific halogen-containing compounds are to be determined. With systems based on vapour phase introduction of sample into the cavity, however, MECA becomes a useful alternative analytical method for halogen compound speciation at sub nanogram levels.

In the case of the metals discussed in this chapter, MECA is unlikely to supplant any of the established analytical methods for their determination. Any methods based on MECA that are developed will likely be indirect, as demonstrated by Rigin (38), who used the technique for determining trace amounts of mercury and other metals in biological and coal-based samples.

REFERENCES

1. B. Gutsche, R. Herrmann, and K. Rüdiger, *Fresenius Z. Anal. Chem.*, **258,** 273 (1972).
2. B. Gutsche and K. Rüdiger, *Fresenius Z. Anal. Chem.*, **297,** 117 (1979).
3. M. Burguera, A. Townshend, and S. L. Bogdanski, *Anal. Chim. Acta*, **117,** 247 (1980).
4. M. Burguera, S. L. Bogdanski, and A. Townshend, *Anal. Chim. Acta*, **153,** 41 (1983).
5. R. Belcher, S. L. Bogdanski, A. C. Calokerinos, and A. Townshend, *Analyst.*, **102,** 220 (1977).
6. S. A. Ghonaim, PhD Thesis, University of Birmingham, 1974.
7. M. Burguera, PhD Thesis, University of Birmingham, 1979.
8. R. Belcher, M. A. Leonard, and T. S. West, *J. Chem. Soc.*, 3577 (1959).
9. M. E. Fernandopulle and A. M. G. MacDonald, *Microchem. J.*, **11,** 41 (1966).
10. M. A. Leonard and G. T. Murray, *Analyst.*, **99,** 645 (1974).
11. F. K. Beilstein, *Ber.*, **5,** 620 (1872).
12. P. T. Gilbert, *Anal. Chem.*, **38,** 1920 (1966).
13. R. M. Dagnall, K. C. Thompson, and T. S. West, *Analyst.*, **94,** 643 (1969).
14. C. V. Overfield and J. D. Winefordner, *J. Chromatograph. Sci.*, **8,** 233 (1970).
15. B. Gutsche and R. Herrmann, *Analyst.*, **95,** 805 (1970).
16. D. A. Stiles, *Proc. Soc. Anal. Chem.*, **11,** 141 (1974).
17. R. Belcher, S. L. Bogdanski, Z. M. Kassir, D.A.Stiles, and A. Townshend, *Anal. Lett.*, **7,** 751 (1974).
18. D. R. Narine, M. E. Peach, and D. A. Stiles, *Anal. Lett.*, **9,** 767 (1976).
19. M. H. K. Abdel-Kader, M. E. Peach, and D. A. Stiles, *J. Assoc. Off. Anal. Chem.*, **62,** 114 (1979).
20. M. H. K. Abdel-Kader, M. E. Peach, M. H. T. Ragab, and D. A. Stiles, *Anal. Lett.*, **12**(A13), 1399 (1979).
21. G. Persaud, R. H. Boodhoo, D. R. Budgell, and D. A. Stiles, *Anal. Chim. Acta*, **177,** 247 (1985).
22. W. I. Awad, Y. A. Gawargious, and S. S. M. Hassan, *Mikrochim. Acta*, 852 (1967).
23. B. Malone, *J. Assoc. Off. Anal. Chem.*, **53,** 742 (1970).

24. J. Solomon and J. F. Uthe, *Anal. Chim. Acta*, **73,** 149 (1974).
25. W. Ladrach, F. Van de Craats, and P. Gouverneur, *Anal. Chim. Acta*, **50,** 219 (1970).
26. M. Burguera, J. L. Burguera, and M. Gallignani, *Anal. Chim. Acta*, **138,** 137 (1982).
27. M. Burguera and J. L. Burguera, *Anal. Chim. Acta*, **153,** 53 (1983).
28. O. Osibanjo and S. O. Ajayi, *Anal. Chim. Acta*, **120,** 371 (1980).
29. Z. M. Kassir, A. A. H. Ta'obi, A. T. Al-Samaraie, and T. A. K. Nasser, *Anal. Chim. Acta*, **172,** 323 (1985).
30. O. Osibanjo, A. Bankole, and S. O. Ajayi, *Analyst.*, **114,** 1483 (1989).
31. R. Belcher, S. L. Bogdanski, S. A. Ghonaim, and A. Townshend, *Nature (London)*, **248,** 326 (1974).
32. J. F. Alder, Q. Jin, and R. D. Snook, *Anal. Chim. Acta*, **123,** 329 (1981).
33. R. Ajlic and R. Stupar, *Vest. Slov. Kemi. Drust.*, **33,** 87 (1986).
34. R. Belcher, S. L. Bogdanski, I. H. B. Rix, and A. Townshend, *Anal. Chim. Acta*, **81,** 325 (1976).
35. W. Hayes, *Proc. Phys. Soc. (London)*, **68** A, 1097 (1955).
36. R. Belcher, S. L. Bogdanski, I. H. B. Rix, and A. Townshend, *Anal. Chim. Acta*, **83,** 119 (1976).
37. S. A. Al-Tamrah, A. Z. Al-Zamil, and A. Townshend, *Anal. Chim. Acta*, **143,** 199 (1982).
38. V. I. Rigin, *Zh. Anal. Khim.*, **42,** 1778 (1987).

INDEX

alcohols, MECA determination 83, 84, 163, 164
  oxidation with NAD^+ method 164–168
    ethanol 164, 165
    ispropanol 166
    methanol 166
    n-propanol 166
  periodate oxidation method 163, 164
    2-amino-2-methyl-1-propanol 163
    1-phenyl-1,2-ethanediol 163
aldehydes, MECA determination 162, 163
amines, MECA determination 84, 133, 134
  n-butylamine 134
  di-n-butylamine 134
  diethylamine 134
  di-n-propylamine 134
  n-propylamine 134
  tertiary amine interference 134
amino acids, MECA determination 79, 84, 118–122, 133, 134
  reaction with carbon disulphide 133, 134
  reaction with TNBS 134, 135
    alanine 135
    glycine 133, 135
    histidine 135
    methionine 135
    serine 135
    threonine 135
    valine 135
  silylation and GC 119–122
    aspartic acid 121, 122
    methionine 121, 122
ammonia, MECA determination 24, 137, 141
  Hesse extraction method 142
  interferences 137–139
  sodium hydroxide method 137–139, 141
    in artesian wells 142
    in coke oven liquors 139
    in effluents 139

ammonium ions, MECA determination 85, 137, 141
  Hesse extraction method 141, 142
  interferences 137–139
  sodium hydroxide method 141
    ammonium acetate 137
    ammonium chloride 137, 139
    ammonium nitrate 137
    in fertilizers 138
    in river water 139
    in soil 141, 142
analysis, electron microprobe 13
antimony, MECA determination 60, 84, 96, 99, 100–110, 137
  improvements in technique 129
  in rocks and other minerals 107
  indirect from S_2 emission 109, 110
arsenic, MECA determination 60, 84, 96, 99, 100–110, 137
  arsenic trifluoride 109
  improvements in technique 129
  in copper and its salts 102
  in oxy cavity 107–109
  in NBS orchard leaves 107
  indirect from S_2 emission 109, 110
  using automated system 68
arsine, MECA determination 103, 104
  flow injection analysis method 107
  separation from other hydrides by GC 103–107
automation IUPAC definition 62

Beilstein halogen test 174
bioluminescence 6
boron, MECA determination 50, 96, 99, 110–114, 137
  boron trifluoride method 113, 114
  improvements in technique 129
  solvent extraction method 111, 112
    interferences 111, 112

boron (*cont.*)
- trimethyl borate method 111–113
  - in steel 113
- bromo- and iodo-compounds, MECA determination 60
- bromo-, chloro- and iodo-compounds, MECA determination 174–186
  - addition of halide solution to indium metal 174, 179
    - combustion tube technique 177
    - limited use of t_m for speciation 175, 176
    - Schöniger flask technique 176
  - addition of halide solution to indium nitrate solution 174, 175
  - after GC separation 178, 179
  - copper halide method 180, 182, 183
    - separations based on t_m value 183
    - interferences 183
  - gallium halide method 180, 181, 183
    - interferences 181, 183
    - iodide enhancement procedure 180, 181
    - spectral interferences 180
  - indium halide method 175, 179
    - interferences 175, 177
    - organohalogen compounds in soil 176, 177
    - spectral interference 175
  - vaporization and transfer to cavity 178, 179

cadmium, MECA determination 189, 190
- interferences 190

candoluminescence 1, 2, 17

carbon, MECA determination 131, 132, 162–163
- indirectly from S_2 emission 132, 162–168
  - acetone 163
  - 2-amino-2-methyl-1-propanol 163
  - benzaldehyde 163
  - *p*-chloraminebenzaldehyde 163
  - cyanide 163, 164
  - *p*-dimethylaminobenzaldehyde 163
  - ethanol 164–167
  - formaldehyde 162
  - isopropanol 166
  - methanol 166
  - n-propanol 166
  - *p*-nitrobenzaldehyde 163
  - 1-phenyl-1,2-ethanediol 163
  - sulphite interference 163

chemiluminescence 5
- chemi-excitation 5
- de-excitation 6, 9
- enzymes 6
- flame 6, 8, 13
  - ambient temperature reaction 6
  - cool 6, 125
  - excited species 6
  - incandescence concerns 25
- gas 6
- liquid 6
- mechanisms 5
  - direct 5, 6, 8, 14, 15
  - in sulphur determinations 73, 77
  - indirect 5, 9, 14, 15
- reactions 5, 8, 9
  - tin(II) bromide 9
  - tin(II) chloride 10
- solid 6
- solutions 6
  - 5-amino-2,3-dihydro-1,4-phthalazinedione 6
  - iophine 6
  - luminol 6
  - 2,4,5-triphenylimidazole 6

chlorine compounds, MECA determination after GC separation 178

chlorine determination 172, 174

chromatography, gas (GC) 13, 59–61
- flame photometric, detector 8, 11, 13, 148
  - dual 13
  - in determination of bromine compounds 174
  - in determination of chlorine compounds 174
  - in determination of fluorine compounds 172
  - in determination of phosphorus compounds 13
  - in determination of sulphur compounds 13
  - MECA 43, 59–61, 119–122, 158
  - selection of wavelength 13
  - spectral interferences 13
- nitrogen carrier gas 13
- separation of mixtures prior to MECA determination 33, 89–91, 103–104, 158, 159

chromatography, high performance liquid (HPLC) 59–61

dual detector 13
MECA detector 43, 59–61, 158
separation of mixtures prior to MECA detection 159–162
cobalt, MECA determination 187

emission intensity, dependence on
    temperature 7, 12, 13, 125, 126
    presence of solid surface 13, 47
emission intensity maximum 18
emission intensity, Salet phenomenon 7
emissions, atomic 1
    bismuth 2
    cobalt 188
    iron 188
    sodium 114, 137
emissions, banded 6
emissions, chemiluminescent 5, 111
    blue 2, 7, 91, 115, 118, 123, 126, 137, 149, 154
    bluish-white 100
    green 2, 7, 9, 13, 110, 126, 132, 148
    red 148
    white 114, 118, 120, 132, 137, 172, 173
emissions, molecular 1, 2
    antimony 100
    arsenic 100
    AsO 118
    boron oxide 110
    C_2 120, 132
    CaF 172
    CaO 2
    CH 132
    CN 132
    copper–halogen 174
    CuH 174
    CuO 174
    CuOH 174
    FeO 188
    FeOH 188
    GcCl 123
    HPO 2, 11, 132, 148
    InBr 174
    InCl 174
    InF 172
    indium–halogen 48, 59, 60, 174
    manganese–halogen 187
    metal–halogen 172, 175
    NH 132
    NH_2 132
    NO 132
    NO–O 118, 120, 132, 137
    OH 15
    phosphorus 7
    PO 13, 148
    S_2 2, 7, 10, 11, 13
    selenium 7
    SnBr 126
    SnCl 1, 22, 126
    SnO 1, 126
    sulphur 6, 7, 12, 13
    tellurium 7, 91
    tin compounds 125

flame, background radiation in MECA studies 23, 53, 112, 137
flame chemiluminescence 13
flame colour 112, 137
flame composition
    hydrogen 1, 2, 6–8, 11
    hydrogen–air 12, 90, 91
        species generation 15
        temperature effects 12
        with nitrogen as diluent 154, 155
    hydrogen based 2
        with argon as diluent 12, 81, 150, 156, 180
        with helium as diluent 100
        with nitrogen as diluent 12, 81, 82, 85, 90, 91, 100, 109, 110, 114, 123, 125, 133, 137, 152, 165, 172, 183, 187, 190
    hydrogen diffusion 8–11, 13, 14, 95, 100, 174, 175
        chemiluminescence 15, 16
        concentration profile 9, 22
        molecular emission spectroscopy 12
        optimum temperature for emission 11
        reaction mechanism 8
        species generation 14, 15
        temperature profile 8, 9, 11, 22
        use in sulphur determinations by MECA 72, 90
    hydrogen shielded 13
flame emission region 7, 125
    central core 7, 126
    periphery 7
flame emissions 114
flame technique 2, 7, 11–13
flame temperature 6–9, 11–13, 77, 110, 125, 148, 154, 155, 183
fluorine determination 171, 172

fluorine, MECA determination 172–174
  interferences 173
  arsenic triuoride method 173
  boron triuoride method 173
  silicon tetrafluoride method 173, 174
    in canal water 173
    in toothpaste 173, 174
    metal fluorides 172, 173
    spectral interferences 173

gallium, MECA determination 180, 191
gallium halides, MECA determination 191
gas analysis 3
germanium, MECA determination 60, 96, 99, 105–107, 123, 124, 187
  germanium oxide 124
  in rocks and other minerals 107
  interferences 124
  separation from other hydrides by GC 105–107

halogens, determination by microcoulometry 177
halogens, MECA determination 171–186
  chlorine, bromine and iodine 174–186
  fluorine 172–174

incandescence 24, 25
  dependence on emissivity of material 24
  dependence on wavelength of emitting radiation 24
    problems in MECA determinations 24, 25, 110, 149
indirect, MECA determinations based on S_2 emission 83, 84, 103, 109, 110, 133–135, 162–167
indirect, MECA determinations based on SiO emission 119–122, 172
indium, MECA determination 187, 191
inorganic analyses 1, 2
iodine determination 172, 174
iron, MECA determination 187

lead, MECA determination 187
linearization, technique, electronic 87

manganese, MECA determination 187
MECA 13–38
  comparison with other flame methods 14
MECA, automation 34, 61–69
MECA Cavity

adaptation to automation 34
aluminium 23, 24, 47–49, 128, 137
as GC detector 119–122
background radiation 23, 24
  minimization by apparatus design 23, 45
  reduction of H_2 emission when using Devarda's alloy 143
brass 112
carbon 47, 48, 63, 75, 79, 80, 92, 123, 150, 152, 153, 165, 180, 190, 191
catalytic breakdown of analyte 29, 77
cleaning 2
conditioning 27, 33
copper 48, 183
design and material of construction 23
Duralumin cavity 95, 158–162
emission burner unit 44, 45
filter unit 83
flame containing unit 50, 60, 96, 104–107, 112, 123, 139, 140
flow injection analysis system 107
formation of refractory compounds 47, 149
heat retention and dispersal 29, 33
hydrogen chloride cavity 50
indium-lined cavity 48, 59, 60
influence of cavity angle to horizontal 23, 48, 52, 60, 63–65
influence of cavity position in flame 52
influence of size 23, 46
insulator, ceramic 28
instrumentation and automation 43–70
iron 81
measurement of T_m 30
metallic 48
modified burner used with spinning cavity probe 51
oxy-cavity 28, 50, 99, 100, 108, 109, 111, 112, 114, 120, 123, 126, 127, 132, 142–144, 158, 172, 173, 190, 191
probe and holder unit 45–52
  automated system 65–69, 85
  holder 50–52
    conventional system 50–52
    automated 63–65
    spinning device 51
  gas generation system 52
    automated (Fl-MECA) 65–69
  probe 45–50
    conventional system 46

INDEX

cavity composition 46, 47
cavity size and shape 46, 47
flame parameters 46
introduction of
    precipitates 109
introduction of solids 47
quartz cavity with aluminium
    cup 81, 82
spinning 51
platinum insert 156
quartz-lined 81
silica-lined 47, 48, 51, 52, 75, 76, 81, 82
stainless-steel 72–75, 77, 78, 82, 84, 85, 91, 92, 128, 133, 137, 139, 140, 143, 144, 148–150, 154, 155, 165, 175, 178, 190, 191
tantalum 47
titanium 47, 81
water-cooled 25, 49, 60, 85, 137, 138, 141, 143, 144, 155, 158, 172, 178
zirconium 47
MECA, chromatography applications 3, 59–61, 89–90, 119
MECA, conventional system 17, 28–33, 43
automated analysers 63–65
    coupling to flow injection system 64, 85
flow injection analysers 85
cavity introduction into flame 47
MECA, discovery and history 1–3, 6–8, 45
MECA emission characteristics 14
addition of air to H_2/N_2 flame in sulphur determination 72
addition of oxygen to H_2/N_2 flame with As and Sb 100
AsO from arsenic trifluoride 109, 173
    spectral interferences of boron and tin 109
AsO from arsine 100, 102–107
    spectral interference of tin 103
    using silver nitrate precipitation method 108, 109
BO_2 from 2-ethylhexane-1,3-diol chelate 111, 112
BO_2 from trimethyl borate 111–113
BO_2 from boron trifluoride 113, 114, 173
cadmium atomic 187, 189, 190
chemiluminescent reactions 29, 72, 77
cobalt–halogen 187, 188
conditioning 27, 33

copper–halogen 183
effect of bond strength on intensity 77
effect of matrix elements on organo-sulphur determinations 78
effect of particle size in Devarda's alloy reductions 143
effect of solid residue retention 33, 47
effect on reproducibility of automating the system 61, 63
enhancement of intensity by water cooling 26
gallium–halogen 180, 181, 183, 191
GeCl from germanium solutions 123
GeO from germane 105–107, 123
HNO compared to HPO emission 132
hot cavity emitters 28, 149
HPO from metal phosphates 149, 150, 152, 153
HPO from organophosphorus compounds 154
    after GC separation 158, 159
    after LC separation 159–161
    conditions giving maximum emission intensity 154–156
HPO from phosphoric acid 148, 149, 150
HPO interference from phosphoric acid 92
incandescence 24, 25, 48, 110, 149
indium–halogen 175–178, 191
influence of cavity surface composition 48, 77
influence of cavity surface reflectivity 24, 33, 47
influence of flame gas composition on intensity 45, 72, 92, 112, 134
influence of heating rate on intensity 28
influence of organic solvents on intensity 32, 77
influence of size on intensity 23, 46
influence of temperature on intensity 26, 72, 77, 92
influence of water as solvent on intensity 32, 33
intensity enhancement with oxygen supply 50
iron–halogen 188
lead–halogen 188
liquid nitrogen use to remove background H_2 emission 94
location 26
manganese–halogen 187

MECA emission characteristics (*cont.*)
  metal–halogen 187–189
  NH_2 from nitrogen monoxide entrained in nitrogen 144
  nickel–halogen 188
  NO–O 119, 132
    dependence on oxygen and flame temperature 132
    from acetonitrile solvent in silylation 119
    from ammonia entrained in air 141–143
    from ammonia entrained in air and nitrogen 141
    from ammonia entrained in nitrogen 143
    from ammonia entrained in oxygen 137, 138, 140, 173
    from nitrogen monoxide entrained in nitrogen 144
    overlap with HPO emission 150
  oxide emitters 99
  oxy-cavity 50, 99, 100, 111–114, 120, 123, 126, 127, 132, 142, 143, 144, 158, 172, 173, 190, 191
  profiles as function of temperature 30
  S_2 from inorganic compounds using conventional system 74–76, 79, 81–85, 92, 135, 149, 150, 153
    destruction with oxygen 50
    effect of mode of breakdown on t_m value 75, 76
  S_2 from inorganic compounds by gas generation 85–88, 115
  S_2 from organic sulphur compounds 76–81, 83, 84, 89, 90, 109, 110, 133, 134, 162–168
    catalytic effect of steel surface 77
    effect of bond strength on emission intensity 77
    effect of matrix components on emission intensity 78, 80
    effect of solute volatility on sensitivity 77
    effect of t_m on sensitivity 77
    interference from sulphuric acid 92
  SbO from stibine 100–107
  Se_2 from inorganic compounds using conventional system 91–94
    destruction with oxygen 50

  effect of treatment with organic acids 92, 93
  variation of emission intensity 92
Se_2 from inorganic compounds by gas generation 94–96
  use of liquid nitrogen to remove excess hydrogen 94
Se_2 from organic compounds 94
selectivity of determinations 28
self absorption in ammonia determination 141
SiCl 114
SiF 114
SiO 114, 119, 120
  in fluorine determinations 172, 173
SnBr 126
SnCl 126
SnO from stannane 103, 105–107, 126–128
  optimum flame conditions 126
  suppression of emission by metal ions 127
  use of aluminium cavity to improve sensitivity 128
SnO 126
solvent and matrix effects 30–33
  avoidance using carbon or silica-line cavity 33, 47
  avoidance using gas generation system 33
  conditioning with alkali metal chloride 33
  evaporation at same time an analyte 30
  multi-peaked response with silica cavity 33
  silicon dioxide as supporting matrix 81–83
spectral overlap 53
Te_2 from inorganic compounds using conventional system 91–93
  effect of treatment with citric acid 93–94
Te_2 from inorganic compounds by gas generation 94
  use of liquid nitrogen to remove excess hydrogen 94
TeO 92, 93
temperature of cavity when t_m is reached (T_m) 30
thallium atomic emission 189–191

effect of Tl–X bond strength on emission intensity 190
thermal excitation 28
time to reach maximum (t_m value) 20
  dependence on heating rate 28, 29
  effect of catalyst 29, 77
  in halogen determinations 175–177, 183
  in phosphorus determinations 150, 155, 160
  in sulphur determinations 20, 21, 28–30, 33, 73–79, 81, 92
  invalidity in gas generation system 33
  order of appearance of emissions 26, 29
  relationship to physical characteristics of analyte 28
  speciation characteristic 21, 28, 29, 30
  temperature dependence 20, 28, 29
water cooling 26
  improvements in design 138, 143
  increase of decrease in emission intensity 26, 138, 155, 179
  temperature control 128, 138
  use with gas generation system 34, 158
  wavelength changes 26
MECA emissions 18, 35–38
  AsH_3 68, 69
  AsO 100
    arsenic trifluoride 109, 173
    arsine 100, 103–109
  AsO_2 50, 95
  atomic 50
    cadmium 187, 189, 190
    iron 188
    lithium 50
    potassium 30
    sodium 50
    thallium 50, 189–191
  band 36
  BO_2 28, 50, 111–113
    boron 2-ethylhexane-1,3-diol-chelate 111, 112
    boron trifluoride 113, 114, 173
    trimethyl borate 112, 113
  C_2 28, 132
    aldehydes 132
  CdO 190
  CH 28, 132
    aldehydes 132
  CN 28, 132

cobalt–halogen 187, 188
continuous 36
CuBr 183
CuCl 183
CuH 183
CuI 183
CuO 183
CuOH 183
FeCl 188
FeOH 188
GaBr 180–183, 191
GaCl 180–183, 191
GaI 191
GeCl 123
GeH 95
GeO 105–107, 123
HNO 132
HPO 25, 92, 148–150, 152–161
In_2 175, 179
indium–bromine 60, 176–179
indium–chlorine 59, 60, 176–179
indium–halogen 48, 175, 187
indium–iodine 60, 179
lead–halogen 188
lead–oxygen 188
manganese–halogen 187
metal–halogen 48, 187
metal hydroxide 50
metal–oxygen 50
NH 132
NH_2 144, 145
nickel–halogen 132, 188
NO 132
NO–O 120, 132, 137, 139, 141–145, 173
OH 50
organic solvents 112
PO 50
S_2 19, 20, 26, 30, 33, 36, 37, 52, 59, 72–90, 92, 109, 110, 115, 132–135, 148, 149, 153, 162–168
S_8 36
SbO 100, 103–107
SbO_2 50, 95
Se_2 18, 91–96
SeO 93
SiCl 114
SiO 50, 114, 119, 120, 172, 173
SnBr 126
SnCl 22, 23, 126
SnH 95
SnO 50, 95, 103, 105–107, 126–128

MECA emissions (cont.)
 Te$_2$ 91–93
 TeO 92–95
MECA functions 18–21
 automated system 61–69
  advantages over non-automated system 62
  sample introduction 63–69
 conventional system 18–21
  advantages over sample introduction by aspiration 22
  cavity as emission promoter 46
  cavity coating to stimulate emission 48
  comparison with gas generation system 33
  co-precipitation technique 94
  de-excitation and emission 20, 21
  evaporation and decomposition of analyte 18, 19, 77
  evaporation of solvent 18
  excited species generation 19
  factors affecting reproducibility 22, 24, 61, 62
  factors affecting sensitivity 22, 24, 61, 62, 77
  influence of cavity and flame parameters 46
  introduction of precipitates 109
  introduction of solids 47
  liquid nitrogen use to remove excess hydrogen 94
  oxidation of organophosphorus compounds 154
  reducing agents 94, 109
  reductive environment of carbon cavity 47
  sample introduction in automated system 63–65
 gas generation system 21, 33–35, 49, 54, 55
  advantages 34
  as GC detector 119–122
  emission processes 21
  factors affecting reflectivity 24
  resolution of mixture components 33–35, 85–88
   gas chromatographic methods 39, 103–107
   physical methods 35

 precipitation of silicon to remove interferences 119
 temperature profile 21
 water cooling 21, 26, 28, 34, 49, 138, 143, 155, 179
 holder 3, 45–52
 incandescence 2, 24, 25, 48
 position in flame 22, 23
 reductive characteristics 19, 47
 reflectivity 24, 47, 151
 residue formation 33, 47, 133
 Salet phenomenon 25–28
  disadvantages of conventional system 33
  in organic sulphur determinations 77
  in phosphorus determinations 153, 155
  in sulphur determinations 73
  mechanism 28
 solid deposits 33, 47, 133
 surface effects 24, 27, 33
 temperature profile 28
MECA, gas generation system 17, 33–35, 43, 49, 54, 55, 85–89
 automated analysers 65–69, 87–89
MECA, instrumentation 2, 44–56, 57
 commercial 44, 55–61
  conventional system 55, 56
   automated analysers 63–65
   MECA 22 spectrometer 55–57
   gas generation system 56–59
   MECA-VAP 43, 54–59
   MEP-101/DIVAP-201 58, 59
 components 44–55
  AAS and FES instrument modification 44
  emission burner unit 44, 45
  cavity probe and holder unit 45–52
  optical unit-readout system 52–54
 detectors 3, 43
  GC 59–61, 158
  HPLC 59–61, 158
 instrumentation and automation 43–70
 sample introduction 17, 28–35
MECA, sample introduction 17, 28–35
 conventional system 18, 28–33
 automated 63–65, 153
 cavity probe and holder unit 45–62, 109, 159–161
 commercial instrumentation 44, 55–61
 gas generation system 18, 49, 54, 55
  advantages 34

INDEX

automated 65–69
chemical methods 34
  acidic generation 34, 35, 112–117, 172
  alkali generation 35, 137–142
  criteria 34
  hydride generation 34, 123, 126, 127
  miscellaneous 35
  redox 34, 35
gas chromatographic methods 35, 119–122
  advantages of MECA over FPD 35
  separation of organophosphates 158, 159
  silylation of amino acids 119
  using MEP-101 59
  liquid nitrogen trapping of NO 144
MECA-VAP 43, 54, 55
  design and characteristics 57, 58
  MEP-101/DIVAP-201 design 58, 59
  mixing of gases behind cavity 143
physical methods 35
resolution of mixture 33
trapping of chemical interferences 145
use of nitrogen purge gas in NO generation methods 144
use of two traps to prevent foaming problems 139
mechanism, energy transfer 6
mercury, MECA determination 167, 168, 187
  biological samples 192
  coal-based samples 192
  inhibition of alcohol oxidation 167, 168
metal determinations 3
metals, MECA determination 84, 187–191, 192
metalloid determinations 1
metalloids, MECA determination 14, 50

nickel, MECA determination 187
nitrite and nitrate, MECA determination 143–148
  copper metal reduction method 145
  Devarda's alloy reduction method 143, 145
    effect of particle size 143
    interferences 143, 144
    potassium iodide reduction method 144
      in admixture 144, 145
      in meats 145, 146
      interferences 144, 145
      spectral interferences 144
  zinc metal reduction method 144, 145
    interferences 145
  cadmium reductor method 145, 146
    in drinking water 145, 148
    interferences 145
    meat protein removal 145, 146
nitrogen, MECA determination 99, 131, 132–148
  direct methods 135–148
    ammonia and ammonium ions 137–143
      ammonium acetate 137
      ammonium chloride 137, 139
      ammonium nitrate 137
      aqueous ammonia 137
      in agricultural samples 141
      in artesian wells 142
      in coke oven liquors 139
      in effluents 139
      in environmental samples 141
      in fertilizers 138
      in river water 139
      in soil 141
      in tobacco leaves 142, 143
  indirect methods based on S_2 emission 133–135
  Hoskins extraction procedure 142, 143
non-metals, determination 3
non-metals, MECA determination 14

pharmaceuticals, MECA determination 66, 86–88
phenol, MECA determination 179
phosphorus, MECA determination 131, 132, 148–162
  advantages over other methods 148, 153
  after GC separation 158, 159
  after LC separation 159–161
  automated system 64, 152–154
    in fertilizers 153, 154
    phosphorus anions 63, 153
  comparison with other methods 152
  effect of cavity incandescence 149
  effect of cavity surface reflectivity 150, 151
  effect of volatility on sensitivity 153
  effect of water cooling on sensitivity 155
  effect of water cooling on t_m 155

phosphorus, MECA determination (*cont.*)
  inorganic compounds 148–162
    ammonium phosphate 149, 150
    barium hypophosphite 149, 150
    batch procedure 151, 152
    in detergents 152
    in rocks 152
    interferences 149
    lecithin 152
    metal phosphates 150
    sodium hypophosphite 149, 150
    sodium phosphate 149, 150
    sodium phospite 149, 150
    sodium polyphosphate 149, 150
    sulphate interference 149
    tetrasodium pyrophospate 149
    total phosphorus 150
  interfering S_2 emission 150, 154
  low temperatue residue build-up 149
  organic 148, 154–162
    after GC separation 158, 159
    after LC separation 159, 160
    insecticides 157, 158
    triethyl phosphate 155
    trimethyl phosphite 155
    tri-n-butyl phosphate 155
    triphenyl phosphine 155
    triphenyl phosphine oxide 155
  phosphoric acid 149
  t_m values 150, 153–155
platinum 7

Salet phenomenon 7, 10, 11, 25–28, 47
selenium, MECA determination 50, 84, 91–96
  elemental selenium 92
    in rocks and other minerals 107
    in selenium sulphide 94
    in shampoo formulations 94
    in sulphuric acid 94
    in water 94
  inorganic compounds 92
    selenium dioxide 92
    sodium selenate 92
    sodium selenite 92
  organic compounds 94
    co-precipitation technique 94, 95
    interferences 94
    Se(IV) determination 95, 96
    Se(TV) and Se(VI) in admixture 95
silicon, MECA determination 99, 114–122
  improvements in technique 129
  silicon tetrafluoride method 115–119
    as silica in iron ore 119
    interferences 118
    spectral interferences 118
silver, MECA determination 167, 168
silylation mechanism 120
solids 1, 2, 7
  MECA determination 47
spectroscopy
  cool flame 3
stibine, MECA determination 103, 104
  shape of calibration curve 72
  solids 80–83
    automated system method 68, 85
    blast furnace slags 81
    coal 81
    criteria for successful analysis 80
    inorganic sulphates 81, 82
    iron ores 81
    MEP-101 method 58, 59
    pitches and lichens 81
    preparation using aluminium cup 81
    sample transfer device 80
    spinning probe method 52
    sulphite and sulphates 81
    sulphide and sulphur 81
    total amount determined by using automated system 85
    teeth 81
    total amount in fuel oils 87
sulphate, MECA determination 75, 76, 81, 82, 85, 86
  in charcoal 75
  in coal 86
  in distilled water 86
  in drinking water 75
  in dusts 75
  in effluents 76
  in eggs 79
  in oil 86
  in orchard leaves 86
  in polyurethane 86
  in soils 75
  in urine 75
  in vinegar 75
  in wine 76
  with sulphite and thiosulphate 86
sulphide, MECA determination 75, 81, 85
  automated system 87, 88
  in effluents 76

sulphite, MECA determination 75, 76, 81, 85, 87
  automated system 87, 88
  in soft drinks 88
  in wine 76
  stabilized with tetrachloromercurate(II) 87
  with sulphate and thiosulphate 86
sulphur, MECA determination 71–91, 72, 81
  after chromatographic separation 89–91
  analytical reliability 73
  automated system 85
    ammonium thiocyanate 85
    hydrogen sulphide 66
    promethazine 84, 85
    sulphur 84
    sulphur anions 85
    sulphur compounds 68
    sulphur dioxide 66, 87
    thiamine 65, 67, 88
    thioacetamide 88
    thiodiacetic acid 88
    thiosemicarbazide 88
    thiourea 64, 85
    total sulphur 85, 87
  gas generation system method 85–89
  inorganic compounds 73–76
    calibration graphs 74
    comparison with nephelometry 75
    dithionate 76
    dithionite 76
    effect of cations on emission intensity 73
    effect of phosphoric acid addition on emission intensity 74, 75
    peroxodisulphate 75, 86
    silica gel adsorbent for SO_2 82
    speciation 75
  organic compounds 76–79
    acetazolamide 77
    amino acids 79
    biotin, H 79
    carbon disulphide 89
    chondroitin sulphate 81
    cysteine 79, 81
    cystine 79, 81
    effect of matrix components 78
    glutathione 79
    in detergents 79, 80
    in egg protein 79
    in solid tablet medication 80
    in vitamins 79
    isopropyl disulphide 89
    ispropyl sulphide 89
    methionine 79, 81
    2-methylthiophene 89
    penicillamine 79
    perpherazine 77
    promethazine 77, 78
    sulphadiazine 77
    sulphamerazine 77, 78
    sulphates 79
    sulphathiazole 77
    sulphonates 79
    sulphonamides 77
    taurine 79
    thionyl chloride 89
    thiosemicarbazide 87
    triamine, B1 79
    triazines 77
    variation of t_m value 76

tellurium, MECA determination 84, 91–96
  powdered tellurium 92
  tellurium dioxide 92
thallium, MECA determination 189, 190
  interferences 190, 191
  thallium acetate 190
  thallium carbonate 190
thiocyanate, MECA determination 75, 76, 86
  in effluents 76
thiosulphate, MECA determination 75, 86
  in effluents 76
  with sulphate and sulphite 86
tin, MECA determination 69, 96, 99, 105–107, 126–128, 187
  improvements in technique 129
  in rocks and other minerals 107
  in tin(II) compounds 126, 127
  separation from other hydrides by GC 105–107

Vol. 62. **Flow Injection Analysis.** *Second Edition* By J. Ruzicka and E. H. Hansen
Vol. 63. **Applied Electron Spectroscopy for Chemical Analysis.** Edited by Hassan Windawi and Floyd Ho
Vol. 64. **Analytical Aspects of Environmental Chemistry.** Edited by David F. S. Natusch and Philip K. Hopke
Vol. 65. **The Interpretation of Analytical Chemical Data by the Use of Cluster Analysis.** By D. Luc Massart and Leonard Kaufman
Vol. 66. **Solid Phase Biochemistry: Analytical and Synthetic Aspects.** Edited by William H. Scouten
Vol. 67. **An Introduction to Photoelectron Spectroscopy.** By Pradip K. Ghosh
Vol. 68. **Room Temperature Phosphorimetry for Chemical Analysis.** By Tuan Vo-Dinh
Vol. 69. **Potentiometry and Potentiometric Titrations.** By E. P. Serjeant
Vol. 70. **Design and Application of Process Analyzer Systems.** By Paul E. Mix
Vol. 71. **Analysis of Organic and Biological Surfaces.** Edited by Patrick Echlin
Vol. 72. **Small Bore Liquid Chromatography Columns: Their Properties and Uses.** Edited by Raymond P. W. Scott
Vol. 73. **Modern Methods of Particle Size Analysis.** Edited by Howard G. Barth
Vol. 74. **Auger Electron Spectroscopy.** By Michael Thompson, M. D. Baker, Alec Christie, and J. F. Tyson
Vol. 75. **Spot Test Analysis: Clinical, Environmental, Forensic and Geochemical Applications.** By Ervin Jungreis
Vol. 76. **Receptor Modeling in Environmental Chemistry.** By Philip K. Hopke
Vol. 77. **Molecular Luminescence Spectroscopy: Methods and Applications** (*in two parts*). Edited by Stephen G. Schulman
Vol. 78. **Inorganic Chromatographic Analysis.** Edited by John C. MacDonald
Vol. 79. **Analytical Solution Calorimetry.** Edited by J. K. Grime
Vol. 80. **Selected Methods of Trace Metal Analysis: Biological and Environmental Samples.** By Jon C. Vanloon
Vol. 81. **The Analysis of Extraterrestrial Materials.** By Isidore Adler
Vol. 82. **Chemometrics.** By Muhammad A. Sharaf, Deborah L. Illman, and Bruce R. Kowalski
Vol. 83. **Fourier Transform Infrared Spectrometry.** By Peter R. Griffiths and James A. de Haseth
Vol. 84. **Trace Analysis: Spectroscopic Methods for Molecules.** Edited by Gary Christian and James B. Callis
Vol. 85. **Ultratrace Analysis of Pharmaceuticals and Other Compounds of Interest.** Edited by S. Ahuja
Vol. 86. **Secondary Ion Mass Spectrometry: Basic Concepts, Instrumental Aspects, Applications and Trends.** By A. Benninghoven, F. G. Rüdenauer, and H. W. Werner
Vol. 87. **Analytical Applications ot Lasers.** Edited by Edward H. Piepmeier
Vol. 88. **Applied Geochemical Analysis.** By C. O. Ingamells and F. F. Pitard
Vol. 89. **Detectors for Liquid Chromatography.** Edited by Edward S. Yeung
Vol. 90. **Inductively Coupled Plasma Emission Spectroscopy: Part I: Methodology, Instrumentation, and Performance; Part II: Applications and Fundamentals.** Edited by J. M. Boumans
Vol. 91. **Applications of New Mass Spectrometry Techniques in Pesticide Chemistry.** Edited by Joseph Rosen
Vol. 92. **X-Ray Absorption: Principles, Applications, Techniques of EXAFS, SEXAFS, and XANES.** Edited by D. C. Konnigsberger